Oliver Zech

Ionic Liquids in Microemulsions

Oliver Zech

Ionic Liquids in Microemulsions

- a Concept to Extend the Conventional Thermal Stability Range of Microemulsions

Südwestdeutscher Verlag für Hochschulschriften

Impressum/Imprint (nur für Deutschland/ only for Germany)
Bibliografische Information der Deutschen Nationalbibliothek: Die Deutsche Nationalbibliothek verzeichnet diese Publikation in der Deutschen Nationalbibliografie; detaillierte bibliografische Daten sind im Internet über http://dnb.d-nb.de abrufbar.

Alle in diesem Buch genannten Marken und Produktnamen unterliegen warenzeichen-, marken- oder patentrechtlichem Schutz bzw. sind Warenzeichen oder eingetragene Warenzeichen der jeweiligen Inhaber. Die Wiedergabe von Marken, Produktnamen, Gebrauchsnamen, Handelsnamen, Warenbezeichnungen u.s.w. in diesem Werk berechtigt auch ohne besondere Kennzeichnung nicht zu der Annahme, dass solche Namen im Sinne der Warenzeichen- und Markenschutzgesetzgebung als frei zu betrachten wären und daher von jedermann benutzt werden dürften.

Verlag: Südwestdeutscher Verlag für Hochschulschriften Aktiengesellschaft & Co. KG
Dudweiler Landstr. 99, 66123 Saarbrücken, Deutschland
Telefon +49 681 37 20 271-1, Telefax +49 681 37 20 271-0
Email: info@svh-verlag.de
Zugl.: Regensburg, Universität, Dissertation, 2010

Herstellung in Deutschland:
Schaltungsdienst Lange o.H.G., Berlin
Books on Demand GmbH, Norderstedt
Reha GmbH, Saarbrücken
Amazon Distribution GmbH, Leipzig
ISBN: 978-3-8381-1672-3

Imprint (only for USA, GB)
Bibliographic information published by the Deutsche Nationalbibliothek: The Deutsche Nationalbibliothek lists this publication in the Deutsche Nationalbibliografie; detailed bibliographic data are available in the Internet at http://dnb.d-nb.de.

Any brand names and product names mentioned in this book are subject to trademark, brand or patent protection and are trademarks or registered trademarks of their respective holders. The use of brand names, product names, common names, trade names, product descriptions etc. even without a particular marking in this works is in no way to be construed to mean that such names may be regarded as unrestricted in respect of trademark and brand protection legislation and could thus be used by anyone.

Publisher: Südwestdeutscher Verlag für Hochschulschriften Aktiengesellschaft & Co. KG
Dudweiler Landstr. 99, 66123 Saarbrücken, Germany
Phone +49 681 37 20 271-1, Fax +49 681 37 20 271-0
Email: info@svh-verlag.de

Printed in the U.S.A.
Printed in the U.K. by (see last page)
ISBN: 978-3-8381-1672-3

Copyright © 2010 by the author and Südwestdeutscher Verlag für Hochschulschriften Aktiengesellschaft & Co. KG and licensors
All rights reserved. Saarbrücken 2010

Contents

Contents i

Preface vii

Constants and Symbols ix

I. Introduction .... - 1 -

II. Fundamentals .... - 5 -

   1. Ionic Liquids .... - 5 -

      1.1. General aspects and types of ionic liquids .... - 5 -

      1.2. Physicochemical properties .... - 8 -

         1.2.1. Melting points and liquid range of ILs .... - 8 -

         1.2.2. Viscosity and ionic conductivity .... - 9 -

         1.2.3. Solvent properties .... - 11 -

      1.3. Applications .... - 13 -

   2. Microemulsions .... - 15 -

      2.1. Definition .... - 15 -

      2.2. Types of microemulsions and phase behavior .... - 16 -

         2.2.1. Oil-in-water and water-in-oil microemulsions .... - 16 -

         2.2.2. Other structures .... - 18 -

         2.2.3. Winsor phase classification .... - 19 -

         2.2.4. Phase diagrams of microemulsions .... - 21 -

         2.2.5. Microemulsions with non-ionic surfactants .... - 22 -

         2.2.6. Microemulsions with ionic surfactants .... - 26 -

      2.3. Applications .... - 27 -

      2.4. Methods to characterize microemulsions .... - 28 -

         2.4.1. Electrical conductivity .... - 28 -

         2.4.2. Viscosity .... - 34 -

2.4.3. Dynamic light scattering (DLS) .... - 35 -
2.4.4. Small angle scattering (SAS) .... - 39 -
2.4.5. Other methods .... - 49 -

III. Experimental .... - 51 -

1. Chemicals .... - 51 -

2. Synthesis .... - 51 -

2.1. Ethylammonium nitrate (EAN) .... - 51 -
2.2. 1-Butyl-3-methylimidazolium tetrafluoroborate ([bmim][BF$_4$]) .... - 52 -
2.3. 1-Butyl-3-methylimidazolium hexafluorophosphate ([bmim][PF$_6$]) .. - 53 -
2.4. 1-Alkyl-3-methylimidazolium chloride ([C$_n$mim][Cl], n = 12, 14, 16,18) ..... - 54 -
2.5. 1-Ethyl-3-methylimidazolium ethylsulfate ([emim][EtSO$_4$]) .... - 55 -
2.6. 2,5,8,11-Tetraoxatridecan-13-oic acid (TOTOA) .... - 56 -
2.7. TOTOA alkali salts .... - 57 -

3. Experimental methods .... - 60 -

3.1. Analytical methods .... - 60 -
3.2. Electrical conductivity .... - 61 -
3.3. Dynamic light scattering .... - 63 -
3.4. Densities .... - 64 -
3.5. Viscosities .... - 65 -
3.6. Small angle X-Ray scattering (SAXS) .... - 65 -
3.7. Small angle neutron scattering (SANS) .... - 66 -

IV. Results and Discussion .... - 69 -

1. The conductivity of imidazolium-based ionic liquids over a wide temperature range. Variation of the anion .... - 69 -

1.1. Abstract .... - 69 -
1.2. Introduction .... - 69 -
1.3. Synthesis and sample handling .... - 70 -
1.4. Results and discussion .... - 71 -

1.5. Concluding remarks...- 78 -

2. Microemulsions with an ionic liquid surfactant and room temperature ionic liquids as polar phase ....- 79 -

2.1. Introduction ....- 79 -
2.2. Investigations at ambient temperature..- 81 -
   2.2.1. Abstract...- 81 -
   2.2.2. Sample handling and experimental path..- 82 -
   2.2.3. Results and discussion ....- 82 -
      2.2.3.1. Microregions and phase behavior..- 82 -
      2.2.3.2. Conductivity ....- 84 -
      2.2.3.3. Viscosity...- 88 -
      2.2.3.4. Dynamic light scattering (DLS) ....- 89 -
      2.2.3.5. Small angle X-ray scattering (SAXS) ....- 90 -
   2.2.4. Concluding remarks...- 97 -
2.3. Ethylammonium nitrate in high temperature stable microemulsions..............- 99 -
   2.3.1. Abstract...- 99 -
   2.3.2. Sample handling and experimental path..- 99 -
   2.3.3. Results and discussion ....- 100 -
      2.3.3.1. Density..- 100 -
      2.3.3.2. Visual observations ....- 101 -
      2.3.3.3. Solubility of EAN in dodecane...- 101 -
      2.3.3.4. Conductivity ....- 103 -
      2.3.3.5. Dynamic light scattering (DLS) ....- 106 -
      2.3.3.6. Small angle neutron scattering (SANS) ....- 109 -
   2.3.4. Concluding remarks...- 118 -
2.4. The effect of surfactant chain length on the phase behavior of microemulsions containing EAN as polar microenvironment ....- 120 -
   2.4.1. Abstract...- 120 -
   2.4.2. Results and discussion ....- 120 -
      2.4.2.1. Phase diagrams ....- 120 -
      2.4.2.2. Density..- 121 -
      2.4.2.3. Conductivity ....- 122 -

Contents

  2.4.2.4. Viscosity .... - 125 -
  2.4.2.5. Dynamic light scattering .... - 125 -
 2.4.3. Concluding remarks .... - 126 -
2.5. Biodiesel, a sustainable oil, in high temperature stable microemulsions containing a low-toxic room temperature ionic liquid as polar phase .... - 128 -
 2.5.1. Abstract .... - 128 -
 2.5.2. Introduction .... - 128 -
 2.5.3. Results and discussion .... - 129 -
  2.5.3.1. Phase diagram .... - 129 -
  2.5.3.2. Visual observations .... - 130 -
  2.5.3.3. Density .... - 131 -
  2.5.3.4. Conductivity .... - 131 -
  2.5.3.5. Small angle X-Ray scattering (SAXS) .... - 134 -
  2.5.3.6. Viscosity .... - 138 -
 2.5.4. Concluding remarks .... - 138 -
2.6. [bmim][BF_4] in high temperature stable microemulsions .... - 140 -
 2.6.1. Introduction .... - 140 -
 2.6.2. Results and discussion .... - 140 -
  2.6.2.1. Conductivity .... - 140 -
  2.6.2.2. Small angle neutron scattering (SANS) .... - 143 -
 2.6.3. Concluding remarks .... - 144 -

3. **Alkali oligoether carboxylates – a new class of ionic liquids** .... - 146 -

 3.1. Abstract .... - 146 -
 3.2. Introduction .... - 146 -
 3.3. Results and discussion .... - 147 -
  3.3.1. Synthesis .... - 147 -
  3.3.2. Conductivity and viscosity .... - 149 -
  3.3.3. Electrochemical stability .... - 152 -
  3.3.4. Cytotoxity tests .... - 154 -
 3.4. Concluding remarks .... - 155 -

V. Summary .... - 157 -

VI. Appendix .... - 163 -

    1.   List of publications .... - 163 -

    2.   Patent .... - 164 -

VII. Literature Cited .... - 165 -

Preface

The work described in this PhD thesis has been carried out at the Institute of Physical and Theoretical Chemistry, Faculty of Natural Sciences IV, University of Regensburg, between October 2006 and December 2009 under the supervision of Prof. Dr. W. Kunz. This work would not have been possible without the help and support of many people, whom I owe a great dept of gratitude.

First of all I would like to express my sincere gratitude to Prof. Dr. W. Kunz for giving me the opportunity to work independently, for the interesting subject, valuable discussions and for financial support.

Furthermore, I am grateful to Dr. D. Touraud for help concerning the formulation of the microemulsions and the phase diagram interpretation as well as for continuous interest in the progress of the work. Additionally, I want to thank Prof. Dr. R. Buchner for discussions concerning the conductivity data of the neat ionic liquids and for critical reading the corresponding manuscript. I am further grateful to PD Dr. R. Müller, Prof. Dr. A. Pfitzner and Prof. Dr. H. Gores for providing their equipment for the TGA, DSC and CV measurements, respectively.

In particular I would like to thank Dr. I. Grillo for her efforts to find the appropriate equipment for the high temperature SANS measurements and valuable discussions concerning the interpretation of the SANS data. Further, I would like to express my gratitude to Dr. P. Bauduin for performing SAXS measurements in Paris and Marcoule and for help with the data evaluation and interpretation. Moreover, I am grateful to Dr. U. Keiderling and Dr. J. Jestin for their efforts concerning SANS measurements in Berlin and Paris, respectively.

In addition, I would like to thank my lab colleagues Dr. S. Thomaier and A. Kolodzieski for the fruitful collaboration in the field of ionic liquids and colloidal systems thereof. Furthermore, I am grateful to J. Hunger, A. Kolodziejski and R. Klein for critical reading parts of this manuscripts, innovative ideas and suggestions. Moreover, I want to express my gratitude to A. Stoppa for the collaboration in the field of conductivity of neat ionic liquids and to M. Kellermeier, Dr. S. Thomaier, E. Maurer, R. Klein, B. Ramsauer and Dr. C. Schreiner for their help and support in the field of alkali oligoether carboxylates. Particular thanks to E. Maurer for performing the cytotoxicity test and Dr. C. Schreiner for the CV measurements. I am grateful to S. Thomaier, A. Stoppa, J. Hunger and H. Dorfner for the teamwork concerning the construction of the homemade glovebox and the barbecue, respectively.

Many thanks to all staff members for the pleasant atmosphere in the laboratory in both a scientific and private manner. I will especially treasure our hikes in the mountains and the "good" morning coffee which we had every morning, although the taste was non-stop miserable and infuriatingly with an extraordinary uniformity.

Apart from that, I am grateful to my colleague M. Dürr, who unfortunately died much too early, for supporting me during my studies, for his friendship and interest. I will miss his unique humor a lot.

In particular I would like to thank my parents for their mental and financial support.

Finally, I want to offer my heartfelt thanks to my girlfriend Nina for her understanding and for encouraging me on all my paths.

Constants and Symbols

Constants

Elementary charge	e_0	$= 1.6021773 \cdot 10^{-19}$ C
Electric field constant	ε_0	$= 8.854187817 \cdot 10^{-12}$ F m^{-1}
Avogadro's constant	N_A	$= 6.0221367 \cdot 10^{23}$ mol^{-1}
Gas constant	R	$= 8.314510$ J (mol K)$^{-1}$
Boltzmann's constant	k_B	$= 1.380658 \cdot 10^{-23}$ J K^{-1}
Planck's constant	h	$= 6.6260755 \cdot 10^{-34}$ J s

Symbols

T	Temperature / K		λ	Wavelength / m
θ	Temperature / °C		q	Scattering angle / m^{-1}
κ	Conductivity / S m^{-1}		D	Diffusion coefficient / m^2 s^{-1}
ρ	Density / g m^{-3}		$I(q)$	Scattering intensity / m^{-1}
Λ	Molar conductivity / S m^2 mol^{-1}		Σ	Specific surface / m^2 m^{-3}
w	Weight fraction		Q	Invariant / m^{-4}
ϕ	Volume fraction		d	Domain size / m
c	Molarity / mol m^{-3}		ξ	Correlation length / m

I. Introduction

The research field of ionic liquids (ILs), which are defined as salts with a melting point below 100°C, began in 1914 with an observation of Paul Walden, who reported on the room temperature liquid salt ethylammonium nitrate (EAN).[1] At that time, this paper did not prompt any significant attention. Nowadays, papers appear faster than 40 per week underlining the extreme growing interest in this field.[2] ILs are frequently termed "green solvents" or "designer solvents".[3,4,5,6] Moreover, they are often considered as future solvents for catalysis,[7,8] chemical reactions,[9,10] extractions,[11] electrochemical purposes[12,13,14,15] and many other potential applications.[16,17] Of particular interest in this context are room temperature ionic liquids (RTILs). The question arises if these substances really keep what they promise. Are ionic liquids really green?

From a toxicological point of view they are not, since it cannot be neglected that large amounts of organic solvents are necessary for the synthesis of most ILs. Furthermore, many ILs, especially imidazolium and pyridinum ILs show a pronounced cytotoxicity and are not biodegradable at all. ILs are termed "designer solvents" as they can be designed for a specific purpose, for example for the optimization of a specific reaction in order to obtain the maximum yield of the isolated product. Depending on the cation and the anion, ILs can be designed low toxic, or exceptionally poisonous (with anions such as cyanide).

There are about 600 conventional solvents used in industry compared to at least 10^6 possible simple ILs.[18] Consequently, there are 10^{12} binary and 10^{18} ternary possible IL combinations. The reputation of "green solvents" mainly arises from the fact that ILs are non-volatile (at least under standard conditions). Hence, ILs do not create atmospheric pollution that results from the volatility of classic organic solvents. Nevertheless, it should be noted that a low vapor pressure does not render ILs green. Consequently, ILs cannot generally be classified as "green solvents". Depending on the application, the recovery, the sustainability and the possibility to design an ionic liquid for a specific task, some ILs can indeed be considered as green.

Apart from the applications of ILs mentioned above, they have also gained interest in classical colloid and surface chemistry. The formation of amphiphilic association structures in and with ionic liquids, such as micelles, vesicles, microemulsions and liquid crystalline phases has been reviewed recently.[19,20,21] The formation of micelles and liquid crystals in EAN was already investigated in the 1980s by Evans and coworkers.[22,23,24]. The self-aggregation of common ionic and non-ionic surfactants in imidazolium based ILs was also reported.[25,26,27] Anderson *et al.* documented micelle formation of SDS in 1-butyl-3-methylimidazolium chloride ([bmim][Cl]) and of Brij 30 in 1-butyl-3-methylimidazolium hexalfluorophosphate ([bmim][PF$_6$]), respectively.[25] Patrascu *et al.* investigated the aggregation behavior of Poly(ethyleneglycol)-ethers in [bmim][BF$_4$], [bmim][PF$_6$] and 1-butyl-3-methylimidazolium bis(trifluoromethylsulfonyl)amide ([bmim][Tf$_2$N]).[26] It was demonstrated that micellar size can be tuned by changing the nature of the RTIL. Hao *et al.* documented vesicle formation of Zn^{2+} fluorous- and zwitterionic surfactants in [bmim][BF$_4$] and [bmim][PF$_6$], repectively.[27] Greaves *et al.* found various protic ionic liquids (PILs) to promote self-assembly of amphiphiles.[21,28] Thomaier *et al.* observed the formation of aggregates of the surfactant-like ionic liquid 1-hexadecyl-3-methylimidazolium chloride [C$_{16}$mim][Cl] in EAN.[29]

From these investigations concerning micellar aggregates in ILs, some general conclusions can be drawn. Compared to water the values for critical micelle concentrations (cmc) in RTILs are significantly higher. Furthermore, the micelle aggregation numbers in RTILs are smaller than those in water, the area per molecule is als o reduced compared to water.

Apart from studies of micellar aggregates in RTILs, such as shape and size of aggregates of the binary system [C$_{16}$mim][Cl]/EAN described in detail in the PhD thesis of S. Thomaier,[30] the formation of liquid crystalline phases has attracted attention. Recently, Zhao *et al.* presented a comprehensive phase diagram of the binary system [C$_{16}$mim][Cl]/EAN.[31] With increasing [C$_{16}$mim][Cl] concentration aggregates of different morphologies have been observed that show many similarities to those found in aqueous systems. In addition to micellar regions, typical liquid crystalline phases, such as normal hexagonal (H_1), lamellar (L_α) and reverse bicontinuous cubic (V_2) phases have been reported.

Surprisingly, most studies did not benefit from the excellent thermal stability of RTILs compared to water. S. Thomaier demonstrated that aggregates of [C$_{16}$mim][Cl] in EAN and [bmim][BF$_4$], respectively are stable up to at least 150°C.[30] The possibility to form high temperature stable self-assembled structures highlights the major advantage of aggregates formed in room temperature ionic liquids. The wide liquid range of RTILs allows in principle the formulation of either high temperature stable or low temperature stable colloidal systems that cannot be obtained in classical binary surfactant/water mixtures.

Starting from these anterior results, the major aim of this thesis was the study of ternary (or pseudo-ternary) mixtures of RTIL/surfactant(+cosurfactant)/oil. The formation of nonaqueous microemulsions with RTILs has already been described in literature. A comprehensive overview about the state of the art concerning these nonaqueous microemulsions is given in section IV. This thesis is mainly focused on the formulation and characterization of high-temperature stable microemulsions with RTILs as polar phase. Herein, the influence of the nature of the RTIL, of surfactant chain length and the nature of the oil on thermal stability and phase behavior of microemulsions have been studied.

II. Fundamentals

1. Ionic Liquids

There is a still growing interest in ionic liquids (ILs) in general and room temperature ionic liquids (RTILs) in particular because of their fascinating and outstanding properties and their wide range of potential applications. In general, an ionic liquid is a liquid that consists only of ions and exhibits a melting point below 100°C. This convention can be justified by the improvement in the range of applications below this temperature. For several decades ILs were considered as a curiosity that was only of partial interest, in particular for special applications. This point of view has changed during the last two decades, the field of ILs is nowadays growing at a very fast rate and several concrete applications have been developed. Giving a comprehensive overview about the current IL research is almost impossible. In the last decade, more than 8000 papers have been published in the field of ILs.[3] It is remarkable that a review about ILs appears every two-to three days, and papers are appearing faster than 40 per week.

Therefore, it is obvious that the scope of ILs cannot be discussed within ten or twenty pages. In the following section, only a brief overview about ILs, their synthesis strategies, physicochemical properties, applications and potential for the future is given.
For more detailed information the reader is referred to the book "Ionic Liquids in Synthesis" by Wasserscheid and Welton[32] and several reviews[4,21,33,34] which give at least an overview about the current IL research.

1.1. General aspects and types of ionic liquids

The definition of ionic liquids allows distinguishing them from a classical molten salt. A molten salt is mostly a high-melting, highly viscous and very corrosive substance while ionic liquids are already liquid at lower temperatures (<100°C) and exhibit in most cases relatively low viscosities. The first IL, ethanolammonium nitrate, described in literature was discovered in 1888 by Gabriel[35] with a melting point between 52-55°C.[35,36] The first true room temperature ionic liquid, ethylammonium nitrate (EAN) was reported in 1914.[1] This polar, colorless liquid exhibits a melting point of 14°C,[22] is supposed to form three-dimensional hydrogen bond networks[23,24] and has an equal number of donor and acceptor sides. In the late 1970s Osteryoung *et al.* and Wilkes *et al.* prepared chloroaluminate melts liquid at room

1.1. General aspects and types of ionic liquids

temperature. However, the so-called first generation ILs with anions such as [AlCl$_4^-$] did not attract much interest because of their sensibility towards hydrolysis. With the discovery of ILs with hydrolysis stable anions the interest in ILs increased rapidly.[2]

In general, to obatin an IL with a low melting point preferably most of the following conditions should be fullfilled. Both cation and anion should be single charged, otherwise the Coulomb interactions would become too strong. The charge should be uniformly distributed to reduce further the Coulomb interactions. Otherwise, too large ions lead to increasing Van der Waals interactions that increase the melting point. Steric hindrance further reduces the ion contact. A low degree of symmetry prevents a regular crystalline packing.

Conventional ILs typically contain bulky organic cations with a low degree of symmetry such as imidazolium, pyrrolidinium, tetraalkylphosphonium, trialkylsulfonium or quaternary ammonium. These cations hinder the regular packing in a crystal lattice. Consequently, the solid crystalline state becomes energetically less favorable, leading to low melting points.[37] This effect can be enhanced further by the implementation of an anion with a delocalized charge, resulting in decreased interionic interactions.[38] A selection of typical cations and anions of ILs is given in Figure I-1.

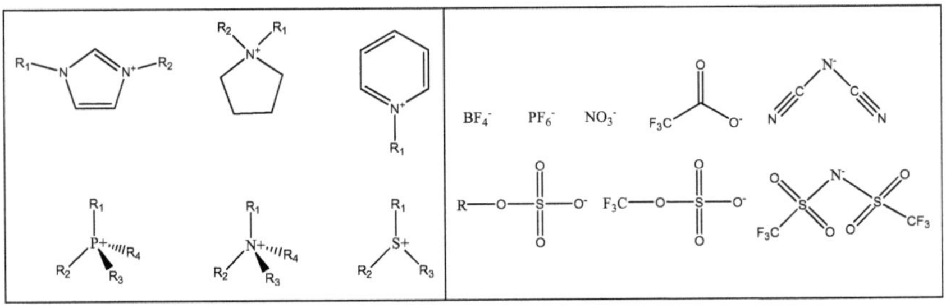

Figure I-1. Some examples for common cations and anions of ILs.

Depending on the combination of the cation and the anion, ILs can have either hydrophilic or hydrophobic character. The most frequently investigated ILs are based on imidazolium cations. Anions such as halides, acetate, nitrate and ethylsufate form hydrophilic ILs while anions such as hexafluorophosphate, bis(trifluoromethylsulfonyl)imid lead to hydrophobic ILs.

Hence, ILs can be either water soluble or not miscible with water.[39,40] However, most ILs are hygroscopic, even hydrophobic ILs assimilate a certain amount of water until saturation is obtained. By the choice and combination of the ions physicochemical properties such as polarity, viscosity, solvation ability, melting point, thermal and electrochemical stability can be targeted. Therefore, ILs are also called "designer solvents" or "task-specific ionic liquids".[33]

Apart from these aprotic imidazolium or pyridinium based ionic liquids there was recently a growing interest in protic ionic liquids (PILs).[21,41] These PILS can be easily obtained by the reaction of a Brønsted acid and a Brønsted base. Ethylammonium nitrate (EAN) is probably the best studied PIL. It is obtained by the reaction of ethylamine with nitric acid in water. The interest in EAN is related to the fact that it exhibits several unique solvent properties associated only with water. These properties infered from the heats, entropies and free energies of solutions of non-polar gases dissolved in EAN.[23] Furthermore, a nearly ideal heat of mixing with water and micelle formation by surfactants dissolved in EAN underline this water-like behavior.[22,24] The interest in ethylammonium nitrate led to the discovery of several protic room temperature ionic liquids, mainly alkylammonium nitrates and thiocyanates. Some of those are summarized in Table I-1.

Table I-1. Physicochemical properties of some selected PILs.

ionic liquid	θ_m / °C	ρ / g·cm^{-3}	η / cP	κ / mS·cm^{-1}
EAN	14[22]	1.216[28]	32[28]	26.9[28]
n-propylammonium nitrate	4.0[41]	1.157[41]	66.6[41]	-
Tri-n-butylammonium nitrate	21.5[41]	0.9176[42]	636.9[42]	-
Dipropylammonium nitrate	5.5[43]	-	-	-
Butylammonium thiocyanate	20.5[43]	0.949[42]	97.1[42]	-
Di-n-propylammonium thiocyanate	5.5[41]	0.964[42]	85.9[42]	-
1-methylpropylammonium thiocyanate	22[43]	1.013[42]	196.3[42]	-

θ_m: melting point; ρ: density at 27°C; η: viscosity at 27°C; κ: specific conductivity at 27 °C

A comprehensive overview about types and physicochemical properties of PILs was currently given in a review of Greaves and Drummond.[21] Ideally, the acid-base reaction in PILs is quantitative so that the only individual species present are the resulting cation and anion. In reality a complete proton transfer from the acid to the base is unlikely and may be less that complete. Hence, both the neutral acid and base species can be present and association of either ions or neutral species can occur. Mac Farlane *et al.* claimed that properties of mixtures of ionic and small amounts of neutral species are clearly of the ionic liquid rather than of the neutral species.[44] The question how high the degree of proton transfer has to be to classify a substance as ionic liquid arises. Mac Farlane *et al.* suggested setting a limit of 1% of neutral species for a substance being defined as pure ionic liquid.[45] Consequently, substances with higher levels of neutral species can be classified as mixtures of ionic liquids and neutral species.

Nevertheless, both aprotic ionic liquids and protic ionic liquids exhibit several outstanding physicochemical properties, the most important ones will be discussed in the following.

1.2. Physicochemical properties

The physical and chemical properties of ionic liquids can be varied by the selection of suitable cations and anions. Therefore, it is possible to optimize the ionic liquid for a specific application. Nevertheless, physicochemical properties are significantly affected by the purity of the substances. Hence, it is indispensable to use very pure starting materials as traces of impurities are hardly to remove, especially if the melting point of the salt is far below room temperature. A second crucial point is the water content of the IL as traces of water also influence the physicochemical properties of the IL. Furthermore, most ILs are hygroscopic, hence, the characterization of ILs should be carried out under inert-gas atmosphere.

1.2.1. Melting points and liquid range of ILs

The key criterion for the evaluation of an ionic liquid is its melting point. It is well known that the characteristic properties vary with the choice of the cation and anion. In this context the relation between structure of cation and anion and melting point is of particular interest. In general, charge, size, symmetry, intermolecular interaction and delocalization of charge are the main factors that influence the melting point.[33,46] With increasing size of the anion the Coulomb interactions in the crystal lattice are weaker and the melting point of the salt decreases. For example, from Cl^- to $[BF_4^-]$ to $[PF_6^-]$ to $[AlCl_4^-]$ the melting points of the sodium salts decrease from 801°C to 185°C. These trends for the sodium salts can be extrapolated to room temperature, the radius of the anion in excess that would be required to obtain a room-temperature liquid sodium salt should be of about (3.4 – 4) Å.[47] In section IV.3., a new room temperature ionic liquid based on a sodium cation will be presented where the properties are not only an effect of anion size, but also of intramolecular charge neutralization through cation complexation. Tetraalkymmonium and phosphonium salts are examples of large cations with delocalized charge. Further, decreasing symmetries in the ions decreases melting points as the ion-ion packing is less efficient in the crystal cell. Additionally, for cations containing alkyl substituents the alkyl chain length has significant effects on the melting point. For example for 1-alkyl-3-methylimidazolium tetrafluroborate salts with increasing alkyl chain lengths, n, the melting points decrease up to $n = 8 - 10$. Beyond this point, the melting points of the salts start to increase again as van der Waals interactions between the hydrocarbon chains contribute to the local structure.[48]

Many ionic liquids have a wide liquid temperature range, frequently from -80°C up to 300°C. For a typical ionic liquid, cooling from the liquid state leads to glass formation at low temperatures. In many cases the glass transition temperatures of ILs are low and lie for 1-alkyl-3-methylimidazolium salts in range between -70 and -90°C.[32] The upper liquid range of ILs is usually related to the thermal decomposition of ILs as most of them are non-volatile. The statement that ILs have no vapor pressure has in the meantime not only theoretically been refuted. In some cases even a distillation of ILs in vacuum was possible.[49,50,51] Protic ILs are in principle readily distillable, as soon as they do not undergo thermal decomposition. The mechanism involves the formation of the original acid and base neutral species by proton transfer and reformation of the PIL on condensing.[49] Destillable PILs require a weakly basic anion, otherwise the IL will undergo thermal decomposition before boiling.[44]

1.2.2. Viscosity and ionic conductivity

In general, ionic liquids are more viscous than most common polar solvents. For most ILs the viscosities, η, at ambient temperature range from 10 cP to 500 cP.[17] For comparison, viscosities of water, ethylene glycol and glycerol at room temperature are (0.89, 16.1 and 943) cP, respectively.[17] The viscosity of ILs is affected by both, their tendency to form hydrogen bonds and the strength of their van der Waals interactions.[40] Within a series of imidazolium based ILs containing the same cation, a change in anion clearly affects the viscosity. The general order $[(CF_3SO_2)_2N^-] \leq [BF_4^-] \leq [CF_3CO_2^-] \leq [CF_3SO_3^-] \leq [(C_2H_5SO_2)_2N^-] < [C_3F_7CO_2^-] < [CH_3CO_2^-] \leq [CH_3SO_3^-] < [C_4F_9SO_3^-]$ with respect to the anion has been found.[32] For ionic liquids with the same anion the trend of increasing viscosity with increasing chain length of

1.2. Physicochemical properties

the alkyl substituent has been found.[32] The viscosities and specific conductivity, κ, of many ILs are strongly dependent on temperature and can be described by the empirical Vogel- Fulcher- Tammann (VFT) equation, particularly for those ILs that show a glass transition temperature. The VFT equation implies the fit parameters A and B and T_0, where the latter is the so called VFT temperature.[52]

$$\ln \eta = \ln A + \frac{B}{T-T_0} \qquad \text{(II-1)}$$

$$\ln \kappa = \ln A - \frac{B}{T-T_0} \qquad \text{(II-2)}$$

The overall trend pyridinium ≤ ammonium < sulfonium ≤ imidazolium in conductivity with respect for the cation has been found.[32] The correlation between type and size of anion and conductivity will be discussed in detail in section IV.1.

The relationship between fluidity and conductance can be considered in terms of the Walden rule,[53]

$$\Lambda \eta = const. \qquad \text{(II-3)}$$

where Λ is the molar conductivity of the ionic liquid defined as $\Lambda = \kappa\, M\, /\, \rho$, where M is the molecular weight and ρ the density of the IL. The Walden rule relates the ionic mobilities which are represented by the molar conductivity to the fluidity of the medium through which the ions move. For a liquid that can well be described by independently moving ions the Walden plot (logΛ versus logη^{-1}) will correspond closely to the ideal line. The ideal line can be represented by using aqueous potassium chloride solutions at high dilutions with a slope of unity.[55]

In other words, in absence of any ion-ion interactions, the slope should be unity. Walden plots of ionic liquids were first described by Angell and coworkers[54,55,56] and MacFarlane and coworkers.[57] Depending on the deviation from the ideal KCl line Angell *et al.* subdivided ILs into different classes, namely superionic liquids, good ionic liquids, poor ionic liquids and non-ionic liquids.[55] This classification is shown in Figure I-2. Substances, whose plot lies more than one order of magnitude below the ideal line, can in this context be classified as "poor" ionic liquids. On that basis, MacFarlane and coworkers demonstrated recently that a number of phosphonium based ILs appear to exhibit strong ion pairing.[57] They proposed to term ILs exhibiting such behavior "liquid ion pairs". PILs can generally be classified as "poor ionic liquids".[55,58]

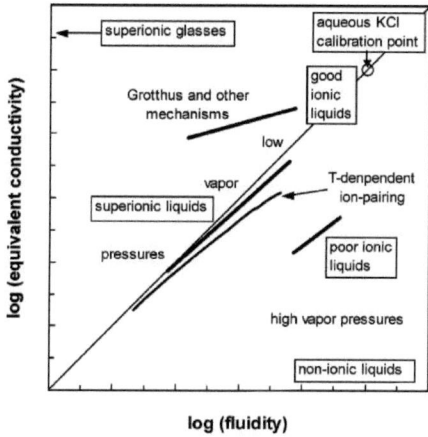

Figure I-2. Classification diagram for ionic liquids, based on the classical Walden rule, and deviations thereof, represented from ref. 55.

1.2.3. Solvent properties

The polarity of a solvent is probably the most widely used solvent classification. One simple qualitative definition is that a polar solvent will dissolve and stabilize dipolar and charged solutes.[17] To evaluate solvent polarities often dielectric constants, dipole moments and refractive indices are used as macroscopic physical solvent polarity parameters. A direct measurement of the dielectric constant which requires a non-conducting medium is not available for ionic liquids. Furthermore, solvent-solvent interactions take place on a molecular level. Therefore, macroscopic physical solvent parameters have often failed in describing solvent effects. In 1965, Reichardt defined solvent polarity as overall solvation capability which depends on the "action of all, nonspecific and specific, intermolecular solute-solvent interactions, excluding such interactions leading to definite chemical alterations of the ions or molecules of the solute."[59] Therefore, empirical solvent polarity scales mostly based on solvatochromic or fluorescent dyes are utilized to classify ionic liquids. The most frequently used dyes to evaluate polarity of ILs are solvatochromic pyridinium N-phenolate betaine dyes.[59,60] An empirical polarity scale called $E_T(30)$ scale has been defined as the molar

1.2. Physicochemical properties

transition energies of the standard betaine dye number 30, measured in different solvents at ambient temperature and pressure

$$E_T(30)/kcal\,mol^{-1} = h\tilde{v}_{max}\,c\,N_A = 28951/(\lambda_{max}/nm) \qquad \text{(II-4)}$$

where $\tilde{v}_{max}$ is the wavenumber and λ_{max} the wavelength of the maximum of the long-wavelength, solvatochromic, intramolecular CT absorption band of the standard betaine dye, and h, c, and N_A are Planck's constant, the speed of light, and Avogadro's constant, respectively.[61] With increasing $E_T(30)$ values the polarity of the substance increases. For nonpolar substances, such as hydrocarbons and perfluorohydrocarbons a more lipophilic betaine dye[62] is used as the standard betaine dye number 30 is not soluble in nonpolar substances. The linear correlation between the E_T values of the two dyes allows the calculation of $E_T(30)$ values for nonpolar substances as well. In 1983 the dimensionless value E_T^N was introduced using water and tetramethylsilane (TMS) as reference solvents to fix the scale with E_T^N (H$_2$O) = 1.00 and E_T^N (TMS) = 0.00.[62] $E_T(30)$ and E_T^N values are known for more than 360 solvents and mixtures thereof.[61] The polarity scale of several organic solvents including different groups of ionic liquids is illustrated in Figure I-3.

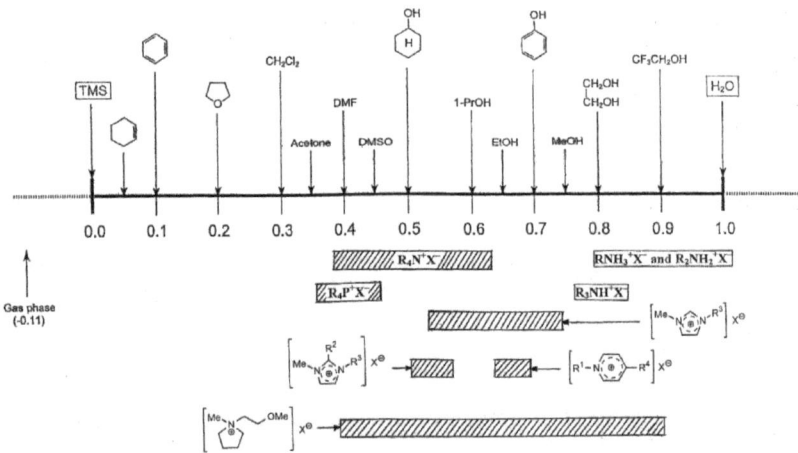

Figure I-3. Normalized solvent polarity scale for several organic solvents and different groups of ionic liquids, reproduced from ref. 61.

- 12 -

For imidazolium based ILs solvent properties, especially the solubility in water are significantly affected by the anion. For example, the [PF$_6^-$] and [NTf$_2^-$] anion lead to water insoluble hydrophobic ILs, anions such as [CH$_3$CO$_2^-$] and [NO$_3^-$] to highly water soluble hydrophilic ILs.[3] Some commonly used anions with respect to their solubility in water are summarized in Figure I-4.

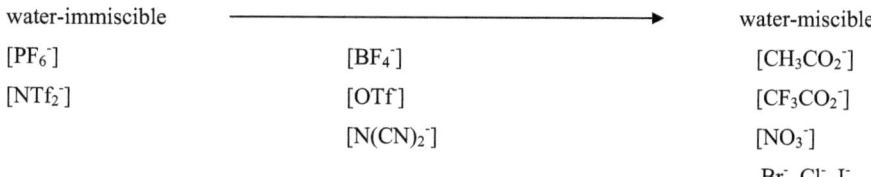

Figure I-4. Effect of some commonly used anions in imidazolium based ionic liquids on their water solubility, redrawn from ref. 3.

1.3. Applications

These outstanding physicochemical properties render ILs, especially room temperature Ionic Liquids (RTILs), excellent candidates for potential applications.[3,4] There are about 600 conventional solvents used in industry compared to at least 10^6 possible simple ILs.[2] Consequently, there are 10^{12} binary and 10^{18} ternary combinations of these possible.

Table I-2. Brief comparison of organic solvents with ionic liquids.

Property	Organic solvents (o.s.)	Ionic liquids
Number of solvents	> 10^3	> 10^6
Flammability	Usually flammable	Usually non flammable
Vapor pressure	Usually remarkable	Negligible
Cost	Normally cheap	2-100 times higher than o.s.
Chirality	Rare	Tuneable
Solvation	Weakly solvating	Strongly solvating
Catalytic ability	Rare	Common and tuneable
Recycability	Green imperative	Economic imperative
Viscosity / mPa s	0.2 - 100	22 - 40000

1.3. Applications

In Table I-2 organic solvents are compared to ILs underlying the most important advantages and disadvantages of ILs. Data were taken from ref. 18 without raising a claim of being comprehensive or representing outliners.

ILs can replace conventional organic solvents in chemical reactions or catalysis[8,7] and can be used as media in extraction processes.[11] Other studies concern the use of ionic liquids as electrolytes in batteries,[12] double layer capacitors or solar cells.[13,14] The electrochemical reactivity in ionic liquids[15] as well as their potential in biocatalysis,[7] catalysis in general,[8] and synthesis[9,10] have been reviewed recently. Furthermore, several industrial applictions have been developed in the last years.[18]

The probably most successful example for an industrial application is the Biphasic Acid Scavenging utilising Ionic Liquid (BASIL^{TM}) process. This process was introduced by BASF in 2002. It is used for the production of alkoxyphenylphosphines, which are generic photoinitiator precursers. Originally, triethyamine was used as acid scavenger yielding triethylammonium chloride as solid waste product. By replacing triethylamine with 1-methylimidazole, the ionic liquid 1-methylimidazolium choride is obatined, which separates from the reaction mixture as discrete phase, the yield increased from 50 % to 98 %. The IL can further by recycled via base decomposition yielding 1-methylimidazole.[63] The process is carried out in a multi-ton scale.

Furthermore, ILs are used as additives in paints, for improved finish and drying processes.[64] Several other industrial applications of ILs, such as dye-sensitised solar cells, lubricant formulations, and additives in lithium ion batteries are in progress.[18]

2. Microemulsions

A typical literature search using the keyword "microemulsion" would return more than 1000 references per year in the past decades. Numerous books and reviews appeared concerning the structure of microemulsions, their phase behavior and applications, which will be discussed in section II.2.3. Hence, it is obvious that everything known about microemulsions cannot be said within a reasonable number of pages. In the following a selection has been made describing only the most important facts and concepts having an intersection with the topic of the thesis.

2.1. Definition

Microemulsions are thermodynamically stable, isotropic transparent mixtures of at least a hydrophilic, a hydrophobic, and an amphiphilic component. In common microemulsions, the polar liquid is water or a brine solution.

The first microemulsions to be recognized as something different from other known structures were described in 1943 by Hoar and Schulman.[65] Herein, structures at that time named "oleophatic hydro-micelle" are nowadays structures called reverse microemulsions. The term microemulsion was first used by Schulman and coworkers in 1959 describing optically isotropic transparent solution consisting of water, oil, surfactant and alcohol.[66] A more recent definition was given by Danielsson and Lindmann:[67] "A microemulsion is a system of water, oil and an amphiphile which is a single optically isotropic and thermodynamically stable liquid solution". The term "water" corresponds to a polar phase that is classically an aqueous solution that can contain electrolytes and other additives. The word "amphiphile" from amphi (both sides) and philos (liking) was coined by Winsor[68] to describe substances with an affinity towards both non-polar and polar phases. Herein, surfactants are the most important amphiphiles. Their amphiphilc character is strong enough to be driven to the interface where the polar part is located in the polar phase and vice versa. The term "oil" refers to an organic phase that is immiscible or at least partially miscible with the polar phase. Therefore, non-polar substances such as hydrocarbons, partially or totally chlorinated or fluorinated hydrocarbons, single-chain alkanes, cyclic or aromatic hydrocarbons, but also triglyceride natural oils can be used.[69] n-Alkanes are the most frequently used non-polar phases in microemulsions.

II.2.2. Types of microemulsions and phase behavior

It has been shown that the polar phase is not necessarily water and the non-polar phase not compulsorily oil. Attempts have been made to formulate and characterize water-free microemulsions,[70,71,72,73] for example by replacing water with ethylene glycerol[74], glycol or formamide.[75] In recent years, progress has been made by formulating nonaqueous microemulsions with Ionic Liquids. An overview about the current state of research concerning ILs in microemulsions is discussed in detail in section IV.2.

2.2. Types of microemulsions and phase behavior

It is well known since hundreds of years that oil and water do not mix. If energy is added to an oil/water system, for example in the form of stirring, very unstable dispersions are formed which phase separate quickly if the system is allowed to relax. The addition of a surfactant reduces the interfacial tension between the two immiscible liquids and a dispersion is formed. Depending on the proportions of the ingredients, either oil-in-water (o/w) or a water-in-oil (w/o) dispersions are formed. These so called macroemulsions have normally dimensions between (0.2-10) µm,[76] they are turbid and thermodynamically unstable. Nevertheless, macroemulsions remain stable for a considerable length of times as they are kinetically stable.[77] By contrast, microemulsions do in principle not require any mechanical work for their formation. The conventional dimensions of microemulsions range from 3 nm to 10 nm.[76]

2.2.1. Oil-in-water and water-in-oil microemulsions

Varieties of different colloidal structures in microemulsions have been described in literature and will be discussed in the following section. The most clear-cut examples of microemulsions are oil-in-water (o/w) and water-in-oil (w/o) microemulsions. In the case of an o/w microemulsion water is the continuous phase with oil droplets stabilized by surfactant molecules. For non-ionic surfactants and ionic double chain surfactants, such as sodium bis (2-ethylhexyl) sulfosuccinate (AOT) no cosurfactant is necessary. For single-chain ionic surfactants a cosurfactant is needed for the formation of a microemulsion, due to the strong repulsion of the charged surfactant head groups. As cosurfactants often *n*-alcohols are used. The easiest cases in this context are spherical micellar aggregates with oil in the core of the swollen micelle as illustrated for an o/w microemulsion in the presence of cosurfactant in Figure II-1. These water continuous microemulsions are often abbreviated as L_1-phases and are still subject of various investigations. This is related to the fact that many applications concern the problem of solubilisation of oil in water.

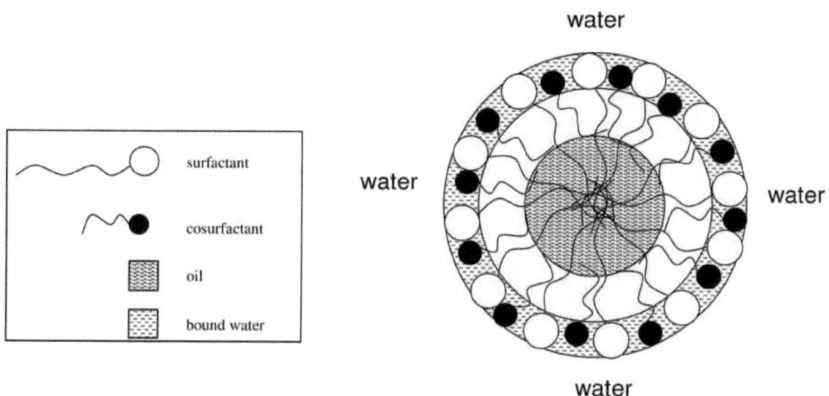

Figure II-1. Schematic illustration of a o/w microemulsion, with oil in the core of the swollen micellar structure.

Conversely, in a w/o microemulsion, oil is the continuous phase with dispersed water droplets stabilized by surfactant molecules. Exemplarily, such w/o microemulsions are illustrated in Figure II- 2 for a spherical aggregate with surfactant and cosurfactant.

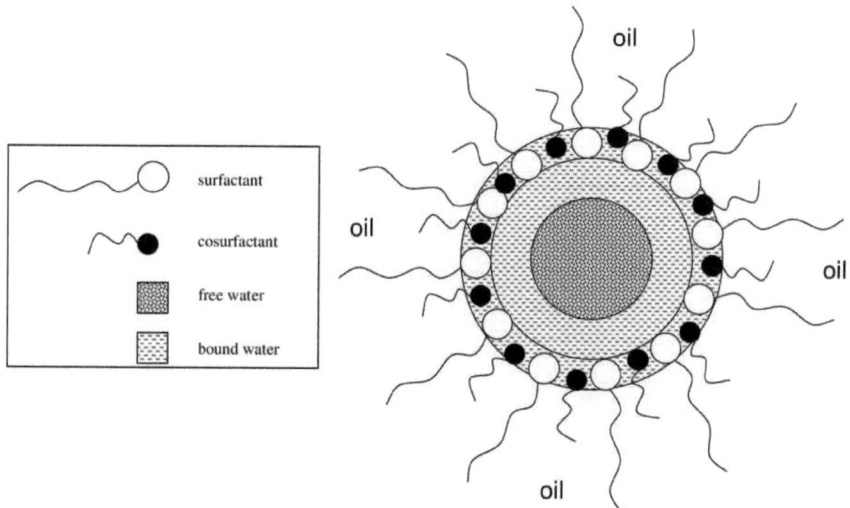

Figure II- 2. Schematic illustration of w/o microemulsions, with free water in the core of the swollen reverse micellar structure and layers of rigidly held water.

II.2.2. Types of microemulsions and phase behavior

The common abbreviation for these oil-continuous structures is L_2-phase. Compared to studies of the physicochemical properties of L_1-phases investigations are scarce on L_2-phases.[78] The L_2-phases have attracted interest in the last years as media for the synthesis of inorganic nanoparticles.[79,80]

The structure of the microemulsion depends on the volume fraction of oil, water and amphiphile as well as on the nature of the interfacial film. o/w microemulsion droplets usually form when the oil volume fraction is low, w/o microemulsions preferentially form when the oil volume fraction is high.

2.2.2. Other structures

In the phase diagram of microemulsions regions exist, where the structure cannot simply be pictured as spherical aggregates depending on the relative ratio of the constituting components. In these regions of the phase diagram several structures have been reported such as rod-like and bicontinuous structures as well as aggregates formed by droplet clusters. Furthermore, several liquid crystalline phases have been found. Bicontinuous structures can be found at almost equal amounts of water and oil. Bicontinuous structures are networks of oil and water nanodomains separated and stabilized by a surfactant interfacial film with a net curvature close to zero. A three-dimensional picture of such a bicontinuous structure is shown in Figure II- 3.[81]

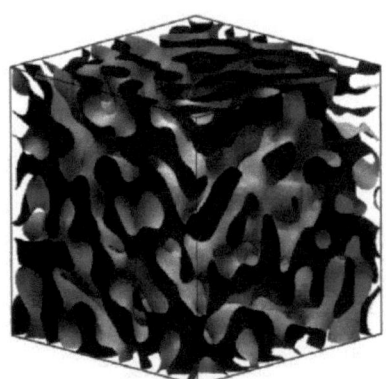

Figure II- 3. Three-dimensional image of a bicontinuous structure taken from ref. 81.

L_3- or sponge phases are isotropic solutions consisting of multiply connected three-dimensional bi-layers.[82] Currently, there is a growing interest in the local dynamics of bi-layers in L_3 phases.[83,84]

Lamellar phases (L_α-phases) are not strictly microemulsions, but they are strongly related to them. They have currently been reviewed by Dubois and Zemb.[85]

2.2.3. Winsor phase classification

One well-known classification of microemulsions was introduced by Winsor[86] who found four general types of phase equilibria:

- Type I: In a type I Winsor system, two phases are in equilibrium. A L_1 type of aqueous micellar system and its extension to an o/w microemulsion is in equilibrium with almost pure oil in the upper phase. Therefore, the surfactant rich phase coexists with the oil phase with small non- aggregated amounts of surfactant. This phase behavior has also often been labelled $\underline{2}$, since it appears as two phases with a surfactant rich lower phase.

- Type II: Contrary, in a type II system, the lower surfactant poor aqueous phase is in equilibrium with an inverse micellar L_2 phase or a w/o microemulsion where oil is the continuous phase. Therefore, the main amount of surfactant is present in the oil phase. Winsor II systems are often noted $\overline{2}$, since the upper phase is the surfactant rich phase.

- Type III: Winsor Type III phases represent a three-phase system, separated into a surfactant-rich phase in the middle and two excess phases, the lower phase composed of almost pure water and an upper oil phase, both with small amount of surfactant. The so-called middle phase microemulsion is in equilibrium with both excess phases and can neither be diluted with water nor with oil. Moreover, this phase is bicontinuous as neither water nor oil is the continuous phase.

- Type IV: This phase represents a true single phase oil, water and surfactant are homogeneously mixed according to Schulman´s definition of microemulsions.[65]

II.2.2. Types of microemulsions and phase behavior

The different Winsor phases are illustrated in Figure II-4.

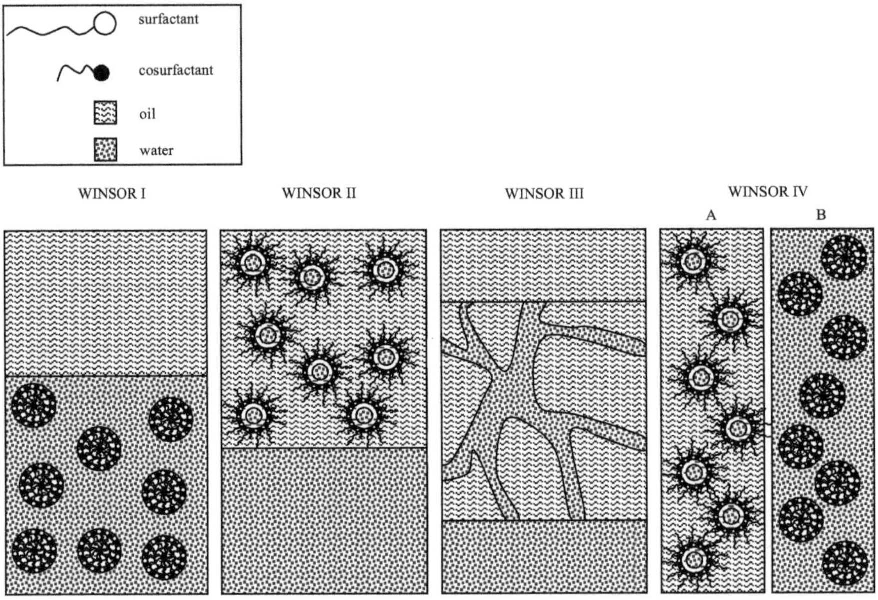

Figure II-4. Different phase forming situations for water-amphiphile-oil mixtures: Winsor phases.

Depending on the type of surfactant and the sample composition the types I-IV form preferentially. Nevertheless, phase transitions can be induced by several parameters such as temperature for nonionic surfactants and electrolyte concentration for ionic ones. For non-ionic surfactants, increasing temperature induces phase transitions from Winsor I to Winsor III to Winsor II. The same order is given for ionic surfactants when the electrolyte concentration increases. Qualitative effects on several variables on changes in phase behavior were summarized by Bellocq et al. for anionic surfactants.[87] These effects are summarized in Table II-1.

Ethoxylated alcohols are the most widely used non-ionic surfactants, they are prepared by the reaction of ethylene oxide with aliphatic or aromatic alcohols.[88] A common abbreviation for these surfactants is C_iE_j where i denotes the carbon number in the hydrophobic tail and j the repetition unit of oxyethylene groups:

$C_iE_j = CH_3(CH_2)_i\text{-}O(CH_2CH_2O)_jH$

Table II-1. Qualitative effect of several variables on the observed phase behavior of anionic surfactants, according to Bellocq et al.[87] Roman numerals denote the different Winsor phases.

Scanned variables (increase)	Ternary diagram transition
Salinity	I → III → II
Oil: Alkane carbon number	II → III → I
Alcohol: low molecular weight	II → III → I
Alcohol: high molecular weight	I → III → II
Surfactant: lipophilic chain length	I → III → II
Temperature	II → III → I

Most of the commercially available C_iE_j surfactants contain a relatively broad distribution of degrees of ethoxylation.[89] In Table II-2 qualitative effects on different variables influencing the phase behavior of ternary systems consisting of non-ionic surfactants (C_iE_j), water and oil are summarized.

Table II-2. Qualitative effect of several variables on the observed phase behavior of non-ionic surfactants, according to Wormuth et al.[90] Roman numerals denote the different Winsor phases.

Scanned variables (increase)	Ternary diagram transition
Salinity	I → III → II
Oil: Alkane carbon number	II → III → I
Pressure	II → III → I
Surfactant: lipophilic chain length	I → III → II
Temperature	I → III → II

2.2.4. Phase diagrams of microemulsions

Maps of the phase behavior of three components systems are often plotted in a "Gibbs triangle", which reflects the ternary composition in two-dimensional space. Exemplarily, a ternary phase diagram of water, oil and amphiphile is shown in Figure II-5.

II.2.2. Types of microemulsions and phase behavior

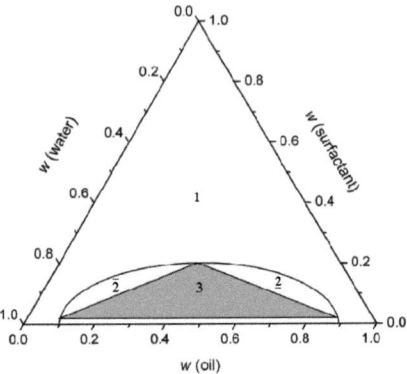

Figure II-5. Schematic phase diagram of water, oil and surfactant mixtures showing regions of 1, 2 and 3 phases.

The compositions of the ingredients may be expressed either in weight, molar or volume fraction. In the majority of cases the compositions are represented in weight fractions, w. Furthermore, the variables temperature and pressure are fixed. Each of the corners of the triangle represents the three pure compounds, while the edges of the triangle map the three binary compositions. In the example given in Figure II-5, the bottom edge denotes water-oil mixtures, the left edge maps oil-surfactant mixtures and the right edge maps oil-surfactant mixtures. The interior of the triangle maps the compositions of all the three components, for example a point exactly at the middle of the triangle denotes equal amounts of water, oil and surfactant (each 33.33 $wt\%$). By moving closer to one of the corners, the composition becomes richer in that particular component. The "1" in Figure II-5 describes the region where a single microemulsion phase occurs, corresponding to a Winsor IV phase. The phase "$\overline{2}$" corresponds to a Winsor II, "$\underline{2}$ " to a Winsor I and "3" corresponds to a Winsor type III phase, respectively.

It is important to distinguish between the phase behavior of microemulsions formed with non-ionic surfactants and ionic ones, as there are many similarities but also some systematic differences.

2.2.5. Microemulsions with non-ionic surfactants

As already mentioned in section II.2.2.1, microemulsions with non-ionic surfactants, especially C_iE_j, can be formed without the addition of any cosurfactant. Basic prerequisite for studying the properties of such microemulsions is knowledge of their phase behavior as it depends on several parameters, such as temperature, pressure and the nature of the components. Probably one of the most important variables in non-ionic microemulsions is temperature,[91] while the effect of pressure has turned out to be rather weak.[92]

The molecular origin of the strong temperature dependence of the phase behavior of C_iE_j containing system is related to the interaction between water and the oligo(ethylene oxide) headgroup of the surfactant.[93] Water can be considered as good solvent for oligo(ethylene oxide) at low temperatures and becomes a bad solvent at more elevated temperatures, resulting in a miscibility gap in the binary water- C_iE_j phase diagram.[94,95] At the so-called cloud point, phase separation occurs into a surfactant rich and a surfactant lean phase. The critical point where the two phases first appear at a specific temperature T_β is called cp_β. Contrary, the miscibility of C_iE_j with oil increases with increasing temperature, the miscibility gap disappears at the critical point cp_α at a defined temperature T_α.

The phase behavior of ternary non-ionic microemulsions has been studied systematically by Kahlweit, Strey and coworkers.[96,97,98] This led to a better understanding of the phase behavior of C_iE_j surfactants, oil and water. Kahlweit et al.[98] presented the temperature dependent phase diagram by vertically stacking the Gibbs triangles into a prism, with temperature on the vertical ordinate. Such a phase prism is shown exemplarily in Figure II-6, reproduced from ref. 90.

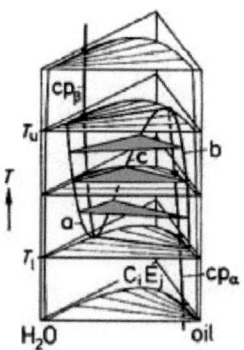

Figure II-6. Schematic phase prism of mixtures of water, oil and non-ionic surfactant as a function of temperature, taken from ref. 90. Tie- lines are shown within the two-phase regions.

II.2.2. Types of microemulsions and phase behavior

The phase prism gives a quite complicated picture of the phase behavior of the ternary system, including critical points, critical endpoints and three phase triangles. At low temperatures the amphiphile is dissolved mainly in the water-rich phase. With increasing temperature, a water-amphiphile nose appears. The three phase triangle appears at a lower critical tie line at a temperature T_l and disappears at an upper critical tie line at a temperature T_u. According to the Winsor phase classification described in section II.2.2.3, the system shows a phase diagram of Winsor type I below T_l. Between T_l and T_u, a Winsor type III system appears and one of type II above T_u can be observed. The position and the width of the three phase temperature interval depend on the nature of surfactant and oil. Generally spoken, for a given oil the three phase temperature interval lies lower, the more hydrophobic the surfactant and for a given amphiphile it lies higher the more hydrophobic the oil.

As these phase prisms are relatively complicated and difficult to interpret, greater understanding comes from going back to two-dimensional space by taking slices through the phase prisms. In these slices it can then be seen how the variables of interest influence the phase behavior. The mass fraction of oil, α, in a two-component water oil mixture is defined as

$$\alpha = \frac{m_O}{m_O + m_W} \qquad \text{(II-5)}$$

and the mass fraction of surfactant, γ, can be expressed as,

$$\gamma = \frac{m_S}{m_S + m_O + m_W} \qquad \text{(II-6)}$$

where m_S, m_O, and m_W are the masses of surfactant, oil and water, respectively.

Figure II-7 represents an idealized pseudo-binary slice through the phase prism obtained for an equal ratio of water to oil ($\alpha = 0.5$). The phase diagram where γ is shown as a function of temperature, θ, takes the shape of a fish. These so-called fish cut is the most often used two-dimensional phase diagram concerning microemulsions with non-ionic surfactants.

The three- phase region (3) lies in the body of the fish, the tail of the fish delimits the single phase region (1) of the fish, and one can distinguish between o/w, bicontinuous and w/o microemulsions depending on the temperature. Furthermore, a $\underline{2}$ region and a $\overline{2}$ region lies below and above the body of the fish, respectively. γ_0 represents the critical microemulsion concentration (cμc). Below this surfactant weight fraction, no mixing of oil and water can be found.

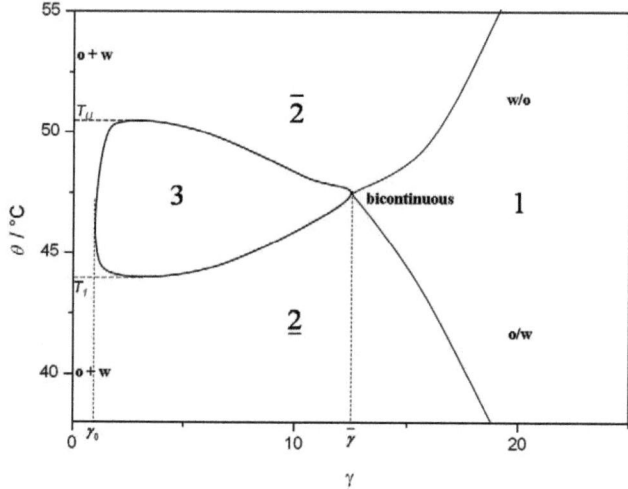

Figure II-7. Fish cut: Schematic phase diagram of equal amounts of oil and water (α = 0.5) as a function of non-ionic surfactant weight fraction and temperature (θ), numbers denote the different phases appearing described in detail in section II.2.2.3. The diagram was redrawn from typical phase diagrams given in literature.[88,90]

As can be seen from Figure II-7, the efficiency, defined as the amount of surfactant that is required to mix equal amounts of water and oil, of an ethoxylated surfactant depends strongly on temperature. The greatest surfactant efficiency can therefore be observed at the point $\bar{\gamma}$ where the body and the tail of the fish meet. It is often called the "optimum formulation" as it is at a given temperature the minimum amount of surfactant that is required to mix equal amounts of water and oil into a single phase microemulsion.

Another useful two-dimensional phase diagram is the χ cut. It is a pseudo-binary phase diagram that is found upon fixing the surfactant weight fraction and varying the ratio of α as a function of temperature. The χ cut will not be discussed here in detail, as the above described fish cut is the most frequently used one. There are many other variables that can influence the phase behavior of ternary systems consisting of water, oil and non-ionic surfactant that are summarized in Table II-2.

II.2.2. Types of microemulsions and phase behavior

2.2.6. Microemulsions with ionic surfactants

As it has been discussed in the previous section for microemulsions with non-ionic surfactants of the C_iE_j type, the hydrophilic lipophilic nature of the surfactant can be tuned with temperature. Hence, it is possible to cover the whole range from o/w to w/o droplets by a simple temperature variation.

Ionic surfactants, such as sodium dodecylsufate (SDS) or dodecytrimethylammoinum bromide (C_{12}TAB) are strongly hydrophilic and do not form large micelles.[99] An upper critical solution temperature in aqueous solution is not observed for ionic surfactants and the size of the micelles does not change significantly with temperature. The hydrophilic-lipophilic balance cannot be tuned simply by temperature variation. Increasing the hydrocarbon tail of ionic surfactants to compensate the high hydrophilicity does not yield microemulsions. Long hydrocarbon tails favour the formation of viscous liquid crystalline phases rather than the formation of microemulsions.[100] Therefore, ionic surfactants do not form microemulsions without the addition of at least a fourth component namely a cosurfactant such as alcohols.

The hydrophobicity can be increased by adding double tails to the surfactant and the tendency of the formation of liquid crystalline phases is reduced. Ionic double-chain surfactants, such as AOT and didodecyldimethylammonium bromide (DDAB) constitute an important class of surfactants and can form three-component microemulsions.[101,102,103,104,105] The major theoretical requirement for the existence of a three-component microemulsion system is that the packing parameter is close to 1.[106] For most of these double-chain surfactants, this condition is fulfilled.[105] Hence, no cosurfactant is necessary for the formation of microemulsions.

On the contrary, for single chain ionic surfactants normally the addition of a cosurfactant is needed in order to form stable microemulsions.[107] The strong electrostatic interaction of the charged surfactant head groups of ionic surfactants makes the addition of cosurfactant necessary to decrease this repulsion and to change the spontaneous curvature.[108] Most commonly low molecular weight aliphatic alcohols are used as cosurfactants, but short chain amines can be used as well.[107,109] The alcohol reduces overall hydrophilicity of the ionic surfactant and favours therefore the formation of microemulsions. The surfactant molecules remain present at the interface between the continuous phase and the dispersed droplet, while the cosurfactant molecules are distributed between oil, water and the interface depending on their solubility. Furthermore, it has been observed that the cosurfactants affect the droplet size, increasing amounts of cosurfactant results in a smaller droplet size. To obtain a middle phase microemulsion, in addition to the presence of a cosurfactant, further addition of salt is required.[110]

The main variables influencing the phase behavior of microemulsions formed with ionic surfactants have been summarized in Table II-1. The addition of salts influences the phase behavior of both ionic and non-ionic microemulsions in the same direction. On the contrary, ionic surfactants respond to temperature in the opposite direction compared to non-ionic surfactants. As already described in section II.2.2.5, the hydrophilicity of non-ionic surfactants increases with increasing temperature, while the water solubility decreases. Ionic surfactants become more water soluble with increasing temperature.[97]

Nevertheless, microemulsions composed of ionic surfactant, water, oil, and cosurfactant and eventually salt tend to be relatively temperature-insensitive. Cosurfactant and salt concentration are the main tuning parameters.[110]

2.3. Applications

The outstanding properties of microemulsions, such as high capacity to solubilize water and oil, low interfacial tension, large interfacial area, spontaneous formation and fine microstructures renders them excellent candidates for a variety of applications that have been summarized in several review papers and books.[90,111,112,113] Here, the probably most important and promising applications described in literature are summarized in order to demonstrate their significance and potential.

The use of microemulsions in enhanced oil recovery (EOR) represents a promising application.[114,115] Approximately 30 % of an oil reservoir can be extracted by primary recovery and another 20 % can be obtained by EOR.[117] Oil remains trapped in the reservoir because of its high interfacial tension of (20-25) $mN \cdot m^{-1}$.[112] A surfactant formulation is injected in the petroleum reservoir, a middle phase microemulsion is formed between excess oil and excess brine and the interfacial tension can be reduced down to 10^{-3} $mN \cdot m^{-1}$. In a similar way microemulsions can be used to remove pollutants from solids[116] and to extract organohalide contaminants.[117]

A second important application is the synthesis of nanoparticles in microemulsions. In this context particularly w/o type microemulsions are of importance. It has been suggested that the

II.2.3. Applications

w/o droplets act as templates and control the nucleation for the particle growth of mainly inorganic nanoparticles.[118,116,119] Both w/o and o/w microemulsions can be used in a similar way to synthesize polymer particles. The obtained polymers exhibit a better defined distribution compared to conventional emulsion polymerization techniques.[111,120]

Since several decades microemulsions are used as lubricants and corrosion inhibitors as the metal surface is protected by a surfactant film. Furthermore, water imparts higher heat capacity to the system compared to pure oil.[112]

Furthermore, microemulsions are applied in drug-delivery systems,[121,122] skin care products[123] and are of industrial importance as they are encountered in agricultural spray formulations.[111] Another interesting point concerns studies of enzymatic reactions in microemulsions. Catalysis and enzymatic activity has been studied over the last years and has been summarized by several authors.[124,125,126]

Other interesting aspects are the use of microemulsions as reaction media in organic reactions and catalyses, in biotechnological applications and in separation science.

2.4. Methods to characterize microemulsions

A variety of methods has been applied to study structure and dynamics of microemulsions and has been summarized in several books and reviews. In the following the most frequently used methods will be discussed. The methods that have been used to characterize the high temperature stable microemulsions of this work will be discussed more detailed than the other ones. Furthermore, it will mainly be focused on characterization techniques of w/o microemulsions, as the microemulsions presented in the thesis mainly concern ionic liquid in oil type microemulsions. Characterization techniques used are hence comparable to w/o- type microemulsions.

2.4.1. Electrical conductivity

Investigations of transport properties of microemulsions, such as electrical conductivity and viscosity provide important information about their internal dynamics. The conductivity of o/w, w/o and bicontinuous microemulsions can be dramatically different and therefore allows a differentiation about the different structures present. For w/o microemulsions, where oil is the continuous phase, the conductance strongly depends on the water volume fraction.

For w/o microemulsions the conductivity is very low as oil is the continuous phase. With increasing amount of water, a rapid increase in electrical conductivity above a certain threshold can be observed in certain cases. This sharp increase continues with the amount of water, sometimes a plateau can be reached where the conductivity does not change significantly with increasing water content. This behavior, especially the sharp increase of several decades in conductivity above a certain threshold is known as percolation behavior and has first been described by Lagues *et al.*[127] In Figure II-8, the specific conductivity versus volume fraction of the dispersed phase for a percolative system is shown. The curve exhibits a sigmoid shape, the corresponding percolation threshold volume fraction, ϕ_P, is marked with a red point on the curve and an arrow on the x-axis.

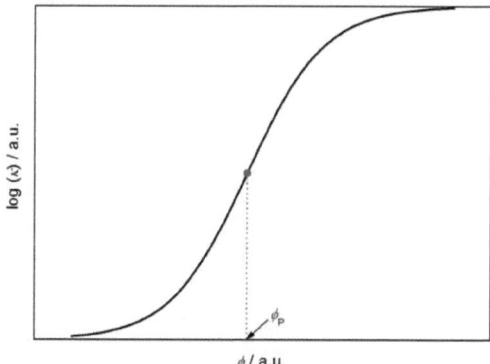

Figure II-8. Idealized evolution of the specific conductivity, κ, versus the volume fraction of the disperse phase, ϕ, for a percolating systems, scale in arbitrary units. ϕ_P denotes the percolation threshold volume fraction.

The phenomenon of percolation in microemulsions has been subject of a large number of investigations and was summarized in several reviews and book chapters. Below the percolation threshold, well separated dispersed water droplets, stabilized by surfactant and eventually cosurfactant are present in a continuous oil medium, the conductivity is very low. With increasing water content the structures start to swell. At the percolation threshold the conductivity changes over several orders of magnitude. One explanation for the sharp increase in conductivity at the percolation threshold implies the formation of droplet associations and clusters, at least the droplets have to come sufficiently close to each other, so that an efficient

II.2.4. Methods to characterize microemulsions

transport of charge carriers can take place. The point of the formation of this random network where long range connectivity is formed is equal to the percolation threshold volume fraction. Below the percolation transition the conductivity is about 10^{-5}-10^{-6} S m^{-1}.[128] Although, the conductivity is very low, this value is still much higher compared to the conductivity of an non-polar solvent which is in the range of (10^{-12}-10^{-16}) S m^{-1}. The conductance of w/o microemulsions at low water volume fractions or of a state much below ϕ_P can be explained following the charge fluctuation model of Eicke et al.[128] In the following this model will be discussed for a w/o microemulsion formed with a cationic surfactant.

The charge fluctuation model assumes spherical droplets with radius r that move independently from each other. In a thermal equilibrium these droplets are in sum uncharged as the number of positive charged surfactant molecules, N_1, is equal to the number of negatively charged of counter ions, N_2. However, due to spontaneous fluctuations, charged droplets will form that carry an excess charge z.

$$z = N_2 - N_1 \qquad (\text{II-7})$$

With the analytical droplet density ρ and the droplet density ρ_i for the different charged droplets i, the mean positive charge, $\langle N_+ \rangle$, and the mean negative charge, $\langle N_- \rangle$, can be expressed as:

$$\langle N_+ \rangle = \sum_i N_1 \frac{\rho_i}{\rho} \qquad (\text{II-8})$$

$$\langle N_- \rangle = \sum_i N_2 \frac{\rho_i}{\rho} \qquad (\text{II-9})$$

As the overall charge of the droplets in sum must be zero due to the electroneutrality condition the mean charge N can be written as:

$$\langle N_+ \rangle = \langle N_- \rangle = N \qquad (\text{II-10})$$

In analogy to diluted electrolyte solutions, the current density j is given by

$$j = \sum_i z_i \, \rho_i \, v_i \, e_0 \qquad (\text{II-11})$$

where v_i is the velocity of migration of the spherical droplets in an electrical field, E. There exists an equilibrium between the electric force and the Stokes friction. With the viscosity, η, the velocity of migration is equal to:

$$v_i = \frac{z_i e_0 E}{6 \pi \eta r} \qquad (\text{II-12})$$

The combination of eq. (II-8) with Ohm's law,

$$\kappa = \frac{j}{E} \qquad (\text{II-13})$$

where κ is the specific conductivity of the microemulsion yields:

$$\kappa = \frac{e_0^2}{6 \pi \eta r} \sum_i z_i^2 \, \rho_i \qquad (\text{II-14})$$

In the case of microemulsion droplets κ can also be expressed in terms of the mean square charge $\langle z^2 \rangle$, which is defined as

$$\langle z^2 \rangle = \sum_i z_i^2 \frac{\rho_i}{\rho} \qquad (\text{II-15})$$

However, due to the electroneutrality the mean charge $\langle z \rangle$ must be equal to zero.

$$\kappa = \frac{\rho e_0^2}{6 \pi \eta r} \langle z^2 \rangle \qquad (\text{II-16})$$

The mean square charge $\langle z^2 \rangle$ can be expressed in terms of the mean-squared fluctuation of the number of ions residing on a droplet $\delta N_i = N_i - \langle N_i \rangle$ with $i = 1, 2$.

II.2.4. Methods to characterize microemulsions

$$\langle z^2 \rangle = \langle \delta N_2^2 \rangle - 2\langle \delta N_2\, \delta N_1 \rangle + \langle \delta N_1^2 \rangle \qquad \text{(II-17)}$$

$\langle \delta N_2\, \delta N_1 \rangle$ can be written as[129]

$$\langle \delta N_2\, \delta N_1 \rangle = k_B T \left(\frac{\partial N_i}{\partial \mu_j} \right)_{p,T,\mu_{k\neq j}} \qquad \text{(II-18)}$$

where k_B is the Boltzmann constant, T the absolute temperature and μ_j the chemical potential of the jth component, with $j = 1, 2$.

The chemical potential μ_j can be expressed as

$$\mu_j = \mu_j^0 + \mu_j^{ex} + k_B T N_i \qquad \text{(II-19)}$$

where μ_j^0 is the standard chemical potential of the ideal solution and μ_j^{ex} denotes the excess chemical potential defined as

$$\mu_j^{ex} = \left(\frac{\partial G^{ex}}{\partial N_i} \right)_{T, N_{k\neq j}} \qquad \text{(II-20)}$$

where G^{ex} is the excess Gibbs free energy. G^{ex} can be set equal to Born's energy of ion solvation, with the dielectric permittivity of the vacuum, ε_0, and the dielectric constant of the solvent, ε_r.

$$G^{ex} = \frac{z^2 e_0^2}{8\pi \varepsilon_0 \varepsilon_r r} = \frac{e_0^2}{8\pi \varepsilon_0 \varepsilon_r r}(N_2 - N_1)^2 \qquad \text{(II-21)}$$

The combination of eq.s (II-14)-(II-17) gives

$$\langle z^2 \rangle = \frac{2N}{1 + 2N \dfrac{e_0^2}{4\pi k_B T \varepsilon_0 \varepsilon_r r}} \qquad \text{(II-22)}$$

From eq. (II-10) and (II-18) the specific conductivity of a dilute microemulsion with the volume fraction of the droplets, ϕ, defined as $\phi = 4 \pi r^3 / 3$ can finally be expressed as

$$\kappa = \frac{\varepsilon_0 \varepsilon_r k_B T}{2\pi\eta} \frac{\phi}{r^3} \qquad \text{(II-23)}$$

It is interesting to note that the final relation between specific conductivity and radius is independent of the charge of the ions.

Another important model to describe specific conductivity of dilute w/o microemulsions below the percolation threshold has been proposed by Kallay et al.[130,131] They distinguish between two different radii, Born's radius r_B and the radius of the water core of the reverse micelle. With thickness of the surfactant layer, l, around the water core, the radius of the whole droplet, r, can be expressed as

$$r = r_B + l \qquad \text{(II-24)}$$

Finally, the following equation is obtained for the specific conductivity of a w/o microemulsion:

$$\kappa = \frac{\varepsilon_0 \varepsilon_r k_B T}{2\pi\eta} \frac{\phi(r-l)}{r^4} \qquad \text{(II-25)}$$

Furthermore, improved charge fluctuation models have been proposed by Hall[132] and by Halle and Björling.[133]

All these models are based on dilute w/o microemulsions below the percolation threshold. In the following more concentrated systems near and above the percolation transition will be considered.

The percolation phenomenon can be divided in two types, static and dynamic percolation. By static percolation a mixture of isolators and solid conductors where the conductivity is almost zero below the threshold is considered. In the dynamic percolation, the droplets are in motion, where the probability for collisions is very high near the threshold condition. Concerning the charge transfer, there are two different mechanisms that can be found in literature. One is in favour with the transfer by hopping of the surfactant from one droplet to another.[134,135] The second one is that the charge transport takes place via ion transfer by interfacial layer opening.[136,137] The model of dynamic percolation[134,138,139] assumes spherical independently moving droplets as in the charge fluctuation model. The conductivity can be described below and above the percolation threshold by appropriate asymptotic power laws, where κ_1 and κ_2 correspond to the conductivities of the percolating droplets and of the homogeneous phase.

$$\ln(\kappa) = \ln(\kappa_1) + \mu \ln\left(\left|\phi - \phi_P\right|\right) \qquad \phi >> \phi_p + \Delta \qquad (\text{II-26})$$

$$\ln(\kappa) = \ln(\kappa_2) - s \ln\left(\left|\phi - \phi_P\right|\right) \qquad \phi << \phi_p - \Delta \qquad (\text{II-27})$$

Δ is the width of the transition region and μ and s are two characteristic exponents and can be determined from the slopes of the equations (II-22) and (II-23)). According to computer simulations the exponents μ and s vary from $\mu = 1.94$ and $s = 1.2$ for dynamic and $\mu \approx 2$ and $s \approx 0.6$-0.7 for static percolation. The dynamic percolation exponent μ was predicted by Grest et al.[134] based on the theory of dynamic percolation and were later confirmed by Kim and Huang,[140] the exponent s was calculated by Derrida et al.[141] Consequently, it is possible to differentiate between dynamic and static percolation of microemulsions. The same scaling laws hold for non-ionic surfactants where the percolation effect can also be induced by temperature variation.[142]

Beside the percolation phenomenon, an antipercolation effect has also been reported.[143,144] In the case of antipercolation the conductivity decreases with increasing water content, salt concentration or temperature. It was suggested that there was no interconnection among the droplets and therefore the probability that a charge hopping transfer takes place decreases.

2.4.2. Viscosity

Viscosity measurements yield important information about the rheological behavior of microemulsions and can show characteristic features. Normally, Winsor I and Winsor II microemulsions are low viscous, Newtonian fluids. Bicontinuous structures often exhibit a non-Newtonian behavior.[142]

A change in the composition of the microemulsion system can affect the viscosity of microemulsions significantly. An oil continuous system can change into a bicontinuous one, followed by a water continuous microemulsion. These structural changes exhibit distinct changes in viscosity. In literature, there is no accurate theory on viscosity of microemulsions, but several semi empirical relations have been found.[145]

Concerning the percolation theory, similar equations as postulated for the conductivity have been suggested.[135] For normal aqueous microemulsions without the addition of any additives (polymers, thickeners), the viscosity percolation phenomenon could not be detected. One condition to observe percolation is that the viscosity of the continuous phase is significantly different from the viscosity of the polar phase. Therefore, it was not possible to verify these equations quantitatively as η_{water} is not sufficiently different from η_{oil}. Boned and coworkers replaced water by glycerol in a microemulsion consisting of AOT, water and isooctane.[74] In that case the condition was fulfilled as $\eta_{oil} \ll \eta_{glycerol}$. They found that the critical exponents obtained from viscosity measurements are very close to those obtained from conductivity measurements.

The conductivity of a well defined and less complex system shows inverse dependence of the viscosity. For example in electrolyte solutions with solvents of different viscosity the rule of Walden is valid (eq. II-3).

In microemulsions it has been found that this rule is often not valid, especially at the stage of percolation. The conductivity increases slowly or rapidly with the volume fraction of the dispersed phase at a constant temperature, the corresponding viscosity may increase or decrease and can pass through a minimum or a maximum.[142,146]

2.4.3. Dynamic light scattering (DLS)

Basically, one can distinguish between two different types of scattering. First, analysis of time dependent fluctuations in the scattered radiation, so-called dynamic- or quasielastic scattering, yields dynamic information. These experiments give information how the particles are moving in Brownian motion and their shapes fluctuate in time. Second, static scattering is described by the measurement of the angle dependence of the scattered intensity and yields structural information. The scattered intensity in absolute scale yields information about the molecular weight of the scattering objects.

Dynamic light scattering has become a standard method for the study of colloidal suspensions

II.2.4. Methods to characterize microemulsions

in general and for the study of microemulsions in particular. In DLS experiments, the time correlation of the scattered light is measured. Therefore, the detected signal depends directly on the motion of the scattering centres in the solution.

Scattered radiation is observed when the investigated system is heterogeneous, for dispersed particles the main scattered signal related to the difference in optical properties, particularly the refractive index difference. The propagation of the incident beam is described by a scattering vector q which is defined as

$$q = \frac{4\pi n}{\lambda} \sin\left(\frac{\theta}{2}\right) \qquad (\text{II-28})$$

where n is the refractive index of the medium (for microemulsions of the continuous phase), λ the laser wavelength and θ the scattering angle with respect to the incident beam (90° for a standard DLS experiment). When a monochromatic beam with the frequency v_0 shines on an immobile particle, the particle emits scattered radiation with the same frequency v_0 of the incident beam in all directions of space. With particles in motion and an immobile observer the moving particle emits scattered radiation of the frequency v_0, but the observer will detect a frequency $v = v_0 + \Delta v$. The frequency Δv is the so-called Doppler shift which depends on q and the velocity of the particle, v. The frequency shift Δv can be positive or negative as the particles are in thermal and random Brownian motion and can hence change direction and speed. The spectrum of scattered particles in Brownian motion looks like a bell shaped curve, the dynamic properties of the scatters can be obtained by the Fourier transform of the spectrum. The latter is the intensity autocorrelation function $g^{(2)}(t)$ which is an average value of the product of the intensity registered at an observation time t, its intensity $I(t)$ and the intensity registered at the time delay τ, $I(t+\tau)$.

$$g^{(2)}(t) = \langle I(t) I(t+\tau) \rangle \qquad (\text{II-29})$$

The intensity autocorrelation function is related to the field autocorrelation function $g^{(1)}(t)$ by the Siegert relation.

$$g^{(2)}(t)=1+B\left|g^{(1)}(t)^2\right|$$ (II-30)

For monodisperse non- interacting spherical particles $g^{(1)}(t)$ exhibits a single exponential decay, where Γ is the decay rate.

$$g^{(1)}(t)=\exp(-\Gamma t)$$ (II-31)

When a baseline A is taken into account, a single exponential decay in terms of the intensity normalized autocorrelation function can be described as:

$$g^{(2)}(t)-1=A+(B\exp(-\Gamma t))^2$$ (II-32)

The decay rate Γ is linked to the translational diffusion coefficient D:

$$\Gamma=Dq^2$$ (II-33)

The diffusion coefficient is related to the hydrodynamic radius by the Stokes- Einstein eq.

$$D=\frac{k_B T}{6\pi\eta R_H}$$ (II-34)

The probably most widely used method for the analysis of quasi elastic light scattering data is the cumulant method. For monodiperse systems $g^{(1)}(t)$ is a mono-exponential decaying function, consequently ln $(g^{(1)}(t))$ as a function of the decay time is a straight line with a slope proportional to the decay rate and inversely proportional to the particle size. In contrast, for polydisperse systems a multi-exponential decay in $g^{(1)}(t)$ can be detected, ln $(g^{(1)}(t))$ as a function of the decay time is no longer a straight line. The cumulant method proceeds by expanding the Laplace transform about the average decay rate $\bar{\Gamma}$ where $C(\Gamma)$ is the distribution function.

$$\bar{\Gamma}=\int_0^\infty \Gamma C(\Gamma)d\Gamma$$ (II-35)

II.2.4. Methods to characterize microemulsions

The results is a series expansion of the exponential function $g^{(1)}(t)$.

$$\ln(g^{(1)}(t)) = \Gamma_0 - \bar{\Gamma}t + \frac{K_2 t^2}{2!} - \frac{K_3 t^3}{3!} + ... \quad \text{(II-36)}$$

A similar fit is obtained by using $\ln(g^{(2)}(t))^{1/2}$ instead of $g^{(1)}(t)$. Nevertheless, light scattering data are relative insensitive to polydispersity. For example one can determine $\bar{\Gamma}$ to better than 1%, K_2 about 20%, but can evaluate only little more than the sign of K_3.[147] The polydispersity index (PDI) is defined as:

$$PDI = \frac{K_2}{\bar{\Gamma}^2} \quad \text{(II-37)}$$

For a PDI < 0.05 a monodisperse, for 0.1 < PDI < 0.2 a narrow and for 0.2 < PDI < 0.5 a broad particle distribution can be assumed.[147]

To get information about the particle distribution nonlinear data analysis can be used. The constrained regularization method (Contin) described by Provencher[148,149] is beside the non-negatively constrained least-squares method (NNLS)[150] the most often used one.[147] The following linear fit model can be used, where $C(\Gamma)$ denotes the decay rate distribution function:

$$g^{(2)}(t) - 1 = \left(\int_{\Gamma_{min}}^{\Gamma_{max}} \exp(-\Gamma t) C(\Gamma) d\Gamma \right)^2 \quad \text{(II-38)}$$

w/o microemulsions are generally very concentrated systems, and far away from the ideal case of a very diluted solution. Consequently, one cannot obtain a free particle diffusion coefficient due to particle interactions and multiple scattering. Moreover, it is well known that the polydispersity in microemulsions is high.[151,152] Hence, for highly concentrated solutions like microemulsions the free particle diffusion coefficient D_0 must be replaced by the so-called effective diffusion coefficient D_{eff}. Consequently, the hydrodynamic radius must be replaced by an apparent hydrodynamic radius R_{Happ}. The interpretation of dynamic light scattering results of microemulsions is difficult. Nevertheless, the R_{Happ} give at least an idea of a dimension and can be compared to other characterization techniques, such as small angle scattering experiments.

2.4.4. Small angle scattering (SAS)

Small angle X-ray scattering (SAXS), small angle neutron scattering (SANS) and light scattering techniques provide the most obvious methods for obtaining quantitative information on size, shape and structure of colloidal particles, since they are based on interactions between incident radiations and particles. Valuable information can be extracted from SAS experiments when the incident wavelength, λ, falls within the size range of the structures to be detected. Therefore, microemulsion droplets can be well characterized by X-rays (λ = 0.5-2.3 Å) and neutrons (λ = 0.1-30 Å).

The overview given here is mainly restricted to static SAS experiments. X-rays are scattered by electrons and neutrons are scattered by nuclei. Hence, each type of radiation has its advantages and disadvantages as a probe for soft matter. The scattering at a single electron is given by its scattering length $b_0^x = 2.8 \cdot 10^{-15}$ m is often called Thomson scattering length. Consequently, the X-ray scattering length of a molecule containing z electrons is given by $b_i^x = z \, b_0^x$. The coherent neutron scattering lengths $b_{i,coh}$ for atoms and isotopes can be found in literature.[153,154] In practise, the mean coherent scattering length density, ρ_{coh}, is a parameter to quantify the scattering efficiency of different components in a system. ρ_{coh} is the sum over all atomic contributions in the molecular volume V_m

$$\rho_{coh} = \frac{\sum_{i=1}^{n} b_{i,coh}}{V_m} = \frac{\rho N_A}{M} \sum_{i=1}^{n} b_{i,coh} \tag{II-39}$$

where $b_{i,coh}$ is the coherent scattering length of the ith atom in the molecule containing n atoms, ρ the density, M the molar mass and N_A is Avogadro's constant. The same equation holds for the calculation of the X-ray scattering length density, ρ_x, when $b_{i,coh}$ is replaced by b_i^x. While X-Ray scattering lengths are proportional to the atomic number, neutron scattering lengths vary with the type of nucleus, which can be positive or negative. A particular

II.2.4. Methods to characterize microemulsions

difference is observed for the coherent scattering length of a hydrogen nucleus ($b_{H,coh}$ = -3.74·10^{-13} cm) and a deuterium nucleus ($b_{D,coh}$ = 6.67·10^{-13} cm).[153] Consequently, a partial deuteration of the components of a sample change the neutron scattering properties completely, while the assumption that the chemical and physical properties do not change significantly has to be made. Nevertheless, the ability to vary the contrast is the main advantage of SANS. Furthermore, neutrons have particular advantage over X-rays for the study of sensitive samples, such as samples of biological material, since X- rays cause molecular degradation due to radiative heating.

The scattering contrast of a particle in solution is defined by the difference of the scattering length of the particle, ρ_P, and the solvent, ρ_S.[155,156]

$$\Delta\rho = \rho_P - \rho_S \qquad (II\text{-}40)$$

The scattering vector, q, is defined as the modulus between the incident wavevector, k_i, and the scattered wavevector, k_f, as shown schematically in Figure II-9.

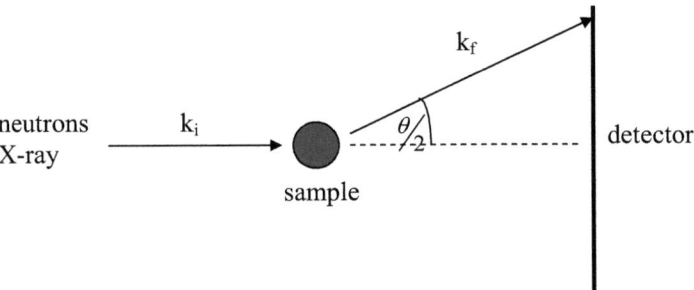

Figure II-9: Schematic correlation between the incident wavevector, k_i, and the scattered wavevector, k_f.

The scattering vector q is related to the scattering angle θ

$$q = |k_f - k_i| = \frac{4\pi n}{\lambda}\sin\left(\frac{\theta}{2}\right) \qquad (II\text{-}41)$$

where λ is the wavelength of the incident beam and n the refraction of the medium.

We consider a coherent beam (neutrons or X- rays) illuminates a sample of the volume V and a thickness d. The incident beam is attenuated on its way through the sample through scattering and absorption. The sample transmission T is defined as the ratio of the outgoing to the incoming intensity, I_i, at a scattering angle $\theta = 0$. μ is the attenuation coefficient. For X-rays, absorption is the essential part of the attenuation and μ is therefore identical to the absorption coefficient. Contrary, for neutrons, the attenuation coefficient can be subdivided in absorption, coherent and incoherent scattering. Coherent scattering means that an incident neutron wave interacts with the nuclei in a sample in a coordinated fashion, the scattered waves can interfere with each other. Incoherent scattering means that the neutron waves interact independently with each nucleus in the sample, the scattered waves cannot interfere with each other.

$$T = \frac{I(\theta = 0)}{I_i} = \exp(-\mu d) \tag{II-42}$$

The radiation with a wavelength λ_i has an incident flux Φ_i, while the beam size is defined by an aperture with an area A. The scattered intensity I is observed at a scattering angle θ and at a distance L within a solid angle $\Delta\Omega$ on the detector. The detector has an efficiency $E(\lambda)$, with $E(\lambda) < 1$. The incident intensity per second I_i can be expressed as:

$$I_i = \Phi_i(\lambda) A E(\lambda) \tag{II-43}$$

The incident radiation interacts with the sample of a number density ρ at a position x ($0 \leq x \leq d$) with the probability $d\sigma/d\Omega$ per unit solid angle. The incident intensity is first attenuated before interacting with the sample by the factor $\exp(-\mu x)$ and further attenuated within the sample with the factor $\exp(-\mu(d-x))/\cos\theta$. The integration over the sample along x yields the intensity $I(\theta)$.

II.2.4. Methods to characterize microemulsions

$$I(\theta) = \int_0^d \Phi_x \, E \, \Delta\Omega \exp(-\mu x) \, A \rho \frac{\exp(-\mu(d-x))}{\cos\theta} \frac{d\sigma}{d\Omega} dx \tag{II-44}$$

As the scattering angles for neutrons and X-rays are sufficiently small, $\cos\theta \approx 1$, the equation can be simplified to:

$$I(\theta) = \Phi_i \, E \, \Delta\Omega \, A \, d \, T \, \rho \frac{d\sigma}{d\Omega} = \Phi_i \, E \, \Delta\Omega \, A \, d \, T \frac{d\Sigma}{d\Omega} \tag{II-45}$$

The first term can be summarized in an instrumental constant $C(\lambda) = \Phi_i \, E \, \Delta\Omega \, A$, while the second term is specific to the sample only.

A complete SAS experiment includes measurements of the scattering intensities of the sample, of its external background (empty cell) and of a standard sample that is necessary for absolute calibration. Furthermore, transmission measurements of the sample, reference sample, empty cell, standard sample and the direct beam have to be performed. A more detailed description concerning the data reduction and absolute calibration is given in the experimental section. As a final result the differential cross section per unit volume $d\Sigma/d\Omega$ is obtained which can be denoted as normalized scattering intensity $I(q)$.

$$I(q) = \frac{d\Sigma}{d\Omega}(q) = \frac{1}{V} \frac{d\sigma}{d\Omega} \tag{II-46}$$

For an ideal diluted solution, each particle is independent from all the others, the scattering intensity $I(q)$ is proportional to the so-called form factor $P(q)$. With increasing concentration, each particle will notice the presence of all the other particles. Therefore, the position of one particle will be dependent of the position of the rest, for example due to particle interaction. This particle interaction can be described by the so-called structure factor $S(q)$. The scattering intensity can be expressed as

$$I(q) = n P(q) S(q) \tag{II-47}$$

where n is the number density $n = N/V$.

Small angle scattering (SAS) experiments have been widely used to study aqueous microemulsions.[92,157,158,159,160,161] Microemulsions are far away from ideal diluted solutions as the surfactant concentration is very high. In most cases SAS spectra of microemulsions exhibit a single broad scattering peak.[159,161,162] Furthermore, a characteristic q^{-4} dependence at large q was observed, which is attributed to a well-defined internal interface.[155,157,163] The peak position varies systematically with surfactant concentration and the water to oil ratio. For the evaluation of scattering data from microemulsions several models have been developed.

Debye model

A prediction of a scattering pattern of a porous solid was already described by Debye *et al.*[164,165] Microemulsions contain a polar and non-polar domains separated by a surfactant film. The system described in the Debye model consists of a continuous matrix with embedded particles, where the scattering regimes are completely independent in their position and shape. The correlation function can be described as

$$g(r) = \exp\left(-\frac{r}{\xi}\right) \tag{II-48}$$

where the loss of "memory" of being located in medium 1 is associated to a translation r from the original point. It was assumed that the decay of the correlation function can be described by one typical size ξ, the so-called correlation length. The corresponding scattering intensity can be written as

$$I(q) = \frac{8\pi \xi^3 \Delta\rho^2}{(1+\xi^2 q^2)^2} \tag{II-49}$$

The Debye model does not comprise a scattering peak. Hence, this model considers only one characteristic length and is no longer applied to describe SAS of microemulsions.

Teubner-Strey model

The probably most popular model was developed by Teubner and Strey.[161] In the Teubner-

II.2.4. Methods to characterize microemulsions

Strey model the structure of the polar and non-polar regimes are not independent from each other.

The oil and water regimes are coupled via the surfactant film resulting in a broad scattering peak. Both, the scattering peak and the q^{-4} dependence at large q can be described with this model. The model is particularly valid for bicontinuous structures. The correlation function $g(r)$ is given by

$$g(r)=\exp\left(-\frac{r}{\xi}\right)\sin\left(\frac{2\pi r}{d}\right)\frac{d}{2\pi r} \tag{II-50}$$

including two parameters, the domain size d, which is a measure of the quasiperiodic repeat distance between polar and nonpolar domains and the correlation length ξ, which can be interpreted as a dispersion of d.[166] Fourier transformation of the correlation function gives the scattering intensity

$$I(q)=\frac{1}{\xi}\frac{8\pi\Delta\rho^2 c_2 V}{a_2+c_1 q^2+c_2 q^4} \tag{II-51}$$

with the scattering volume V and the fitting parameters a_2, c_1, c_2. For the parameters the stability condition $4\,a_2\,c_2 - c_1^2 > 0$ with $a_2 > 0$, $c_1 < 0$, $c_2 > 0$ must be fulfilled. The relation between the two characteristic lengths scales can be expressed as

$$\xi=\left[\frac{1}{2}\left(\frac{a_2}{c_2}\right)^{1/2}+\frac{1}{4}\frac{c_1}{c_2}\right]^{-1/2} \tag{II-52}$$

$$d=2\pi\left[\frac{1}{2}\left(\frac{a_2}{c_2}\right)^{1/2}-\frac{1}{4}\frac{c_1}{c_2}\right]^{-1/2} \tag{II-53}$$

The position of the maximum of the scattering, q_{max}, is related to both, the domain size and the correlation length.

$$q_{max} = \left(\left(\frac{2\pi}{d}\right)^2 - \frac{1}{\xi^2}\right)^{1/2} \tag{II-54}$$

Alternatively the scattering intensity can be expressed as

$$I(q) = \frac{I_0}{(1 - I_0/I_{max})(q^2/q_{max}^2 - 1)^2 + I_0/I_{max}} \tag{II-55}$$

Where $I_{max} = I(q_{max})$ is the maximum scattering intensity and $I_0 = I(0)$ the scattering intensity at the limit $q \to 0$. The ratio d/ξ yields information about the polydispersity of the polar and non-polar domains, the order increases the smaller this ratio.

The amphiphilic factor f_a[167,168,169] can be calculated according to

$$f_a = \frac{c_1}{(4 a_2 c_2)^{1/2}} \tag{II-56}$$

The factor f_a approaches 1 for the disorder line, where the solution loses its quasiperiodical order.[54] The liquid crystalline lamellar phase corresponds to $f_a = -1$.[169] These values delimit the region, where microemulsions may be found.[168] The factor ranges from -0.9 to -0.7 for well-structured bicontinuous microemulsions.[167,169]

Direct determinations from scattering curves

The most important parameters are the specific area Σ at the polar/non-polar interface and the polar and non-polar volume fractions ϕ_{pol} and ϕ_{apol}. The polar and non-polar volume fraction can be calculated from the known composition of the sample and the corresponding molar volumes. The specific area can be obtained with Porod's law.[163] For a sharp interface the scattering intensity $I(q)$ is decreasing with q^{-4} at large q. The Porod law can be expressed as

$$\lim_{q \to \infty}(I q^4) = 2 \pi^2 \Delta\rho^2 \Sigma \tag{II-57}$$

where $\Delta\rho^2$ is the square of the scattering contrast between the polar and the non-polar part.

II.2.4. Methods to characterize microemulsions

For concentrated systems that do not consist of separated particles and even for bicontinuous structures the same equation holds.[170] At least mathematically one can go far enough in q so that a piece of surface is isolated and decorrelated from any other part of the surface. When the polar/non-polar interface is rough, no Porod limit can be observed at large q. A Porod limit is observed when $I\,q^4$ versus q^4 yields a limiting value at large q. When a Porod limit is observed, the experimental invariant Q_{exp} can be determined. From a formalistic point of view the integral goes from zero to infinity, but the q-range is limited. Therefore, an additional term can be added that takes this contribution into account.

$$Q_{exp} = \int_0^\infty I(q)q^2\,dq = \int_0^{q_{exp}} I(q)q^2\,dq + \frac{[I(q)q^4]}{q_{exp}} \tag{II-58}$$

For the determination of Q_{exp} it is essential to subtract the background, due to the incoherent contribution in a neutron scattering experiment for instance. The background A can be obtained from the plot $I(q)q^4$ versus q^4 as $I(q)q^4 = B + A\,q^4$.

Q_{exp} should be identical to the theoretical invariant Q_{theo} which is given by

$$Q_{theo} = 2\,\pi^2\,\Delta\rho^2\,\Phi_{pol}\,(1-\Phi_{pol}) \tag{II-59}$$

The obtained specific surface Σ can be used to evaluate the structure, as different models have been proposed, such as models for bicontinuous structures, w/o spheres and repulsive spheres.

Cubic random cell model (CRC)

The CRC model was first described by Jouffroy and coworkers[171] and is valid for bicontinuous structures. This model assumes that the structure can be described by a set of cubes of the size d^* filled with water or oil. The size is the so-called persistence length and is related to q_{max}.

$$d^* = 2\pi/q_{max} \tag{II-60}$$

The value of d^* can be set equal to the domain size d in the Teubner-Strey model. For cubic random cells that are covered by a surfactant film, the persistence length can be written as

$$\Sigma d = 6 \, \Phi_{pol} \, (1 - \Phi_{pol})$$ (II-61)

The CRC model is valid when $0.18 < \Phi_{pol} < 0.82$.[172]

Spheres of water in oil

For w/o spheres the relation between the specific surface, d and the volume fraction can be expressed as:[173]

$$\Sigma d = 4.84 \, \Phi_{pol}^{1/3}$$ (II-62)

Repulsive spheres

For repulsive spheres the relation between the specific surface, *d* and the volume fraction can be written as:[157]

$$\Sigma d = 4.32 \, \Phi_{pol}^{2/3}$$ (II-63)

Generalized indirect Fourier transformation (GIFT)

The scattering intensity $I(q)$ can be expressed as a product of form factor $P(q)$ and structure factor $S(q)$ as already described in eq. (II-43). Microemulsions are polydisperse systems, in the case of polydisperse interacting particles the scattering intensity can be written in a similar way,

$$I(q) = n \overline{P}(q) S_{eff}(q)$$ (II-64)

where *n* is the number density, $\overline{P}(q)$ is the mean form factor averaged over all particle dimensions and $S_{eff}(q)$ is the effective structure factor of interacting spheres.[174,175,176]

$$\overline{P}(q) = \sum_k x_k \, f_k^2 \, B_k^2(q)$$ (II-65)

II.2.4. Methods to characterize microemulsions

$$S_{eff} = \frac{1}{P(q)} \sum_{k,l} f_k f_l B_k(q) B_l(q) S_{kl}(q) \tag{II-66}$$

where x_k is the mole fraction of species k, f_k the form amplitude of species k at $q = 0$, B_k the normalized form amplitude of k and S_{kl} the partial structure factor describing the hard sphere interaction between species k and l. It is important to note that for polydisperse interacting spheres $S_{eff}(q)$ is not independent of the form factor. Hence, $S_{eff}(q)$ also includes particle properties. Furthermore, the relation between the isothermal compressibility and the structure factor at $q = 0$ is not valid for $S_{eff}(q)$.[177]

Another possibility to evaluate small angle scattering spectra of such concentrated systems is possible with the "generalized indirect Fourier transformation" (GIFT) method.[178,179,180] The GIFT method allows to determine the form factor $P(q)$ and the structure factor $S(q)$ simultaneously, whereby $P(q)$ is extracted in a model free way. Recent applications have shown, that the GIFT method can also be used for polydisperse and nonglobular systems.[167,181] In the case of polydisperse interacting particles, the scattering intensity can be expressed as

$$I(q) = n\, P_{MF}(q)\, S_{eff}(q) \tag{II-67}$$

where P_{MF} is the form factor extracted in a model free way. Furthermore, the pair distance distribution function $p(r)$ is obtained. The $p(r)$ is more descriptive, because real distances within the particles are obtained. Hence, it is possible to make directly a qualitative estimation about the particle shape. The p(r) function is related to the convolution square of the scattering length density.

$$p(r) = r^2 \Delta \tilde{\rho}^2 \tag{II-68}$$

The $p(r)$ function approaches to zero at the maximum particle dimension and represents a histogram of distances inside the particle weightened by the scattering length density differences. Hence, one can extract the diameter from this function. The area under the $p(r)$ function is proportional to the forward-scattering intensity.

The effective structure factor in terms of Percus-Yevick closure relation and hard sphere interaction yields a correlation peak similar to that of the Teubner-Strey model. It is valid for uncharged systems and has often been used to describe SAS curves of microemulsions. It is described by the volume fraction, polydispersity and effective interaction radius R_{eff}.

2.4.5. Other methods

One important method to visualize microemulsion structures is freeze-fracture transmission electron microscopy (FFTEM). Samples are rapidly frozen, fractured and replicated with a thin metal film. The metal replica of the fracture face yields images of the local structure of the microemulsion phase. FFTEM images of microemulsions should be considered with care in order to get not misled by artefacts.[182] It has been shown that FFTEM images of bicontinuous microemulsions are in agreement with knowledge from these systems by other characterization techniques, such as SAS.[92,183,184] More recently, a freeze fracture direct imaging technique has been developed that also allows to take micrographs of oil-rich microemulsions, reliable images of w/o microemulsions could be obtained.[185]

^{1}H-NMR self diffusion measurements have been used to investigate the type of microemulsion, microstructure and phase transitions.[92,186,187,188,189] Especially bicontinuous microemulsion structures can be observed by this method.[189,190] In a typical experiment, the molecular self-diffusion coefficients of each component can be determined simultaneously. Comparison of their diffusion coefficients yields important information about the microemulsion structure. For a o/w droplet structure, surfactant and oil diffusion coefficients are in the same order of magnitude, as they diffuse as droplets. For o/w structure, diffusion coefficient of water is only slightly reduced compared to pure water, as water is the continuous phase. In the case of w/o structures with oil as continuous phase, water and surfactant have the same diffusion coefficients while the diffusion coefficient of the oil phase is very similar to its pure state. On the contrary, for bicontinuous structures diffusion coefficients of water as well as of oil are high, both water and oil diffuse freely within the water and oil domains, respectively. Their diffusion coefficients are only slightly reduced compared to the pure compounds. In a bicontinuous structure the surfactant molecules exhibit a lateral diffusion within the film. Compared to a droplet structure, the lateral diffusion coefficient is significantly higher. Consequently, NMR self-diffusion experiments allow distinguishing between w/o, o/w and bicontinuous structures.

NMR spin relaxation measurements yield information about the rate of molecular reorientation or tumbling. Reorientation involves the rotational diffusion of the droplet and the lateral diffusion of the surfactant within the surfactant film.[191] With a known lateral

II.2.4. Methods to characterize microemulsions

diffusion coefficient, a quantitative analysis of the droplet size can be made. For small droplets, the reorientation is fast, while it is slow for large droplets. However, these kind of studies a scarce in the field of microemulsions as these studies require deuterium-labeled surfactants and the size determination is not as direct as with SAS experiments.[189]

III. Experimental

1. Chemicals

1-chlorobutane (Fluka ≥ 99.5%), ethyl bromide (Fluka, 99%) and 1-butylimidazole were distilled prior to use. 1-chlorododecane (Merck, ≥ 95%), 1-chlorotetradecane (Aldrich ≥ 98%), 1-chlorohexadecane (Fluka ≥ 97%) and 1-chlorooctadecane (Merck ≥ 98%) were used as received. 1-Methylimidazole was distilled from KOH (Merck ≥ 99%), stored over molecular sieves and redistilled prior to use. Aqueous solutions of ethylamine (Fluka, 70%) and nitric acid (Merck, 65%), as well as sodium tetrafluoroborate (Aldrich, 98%), potassium hexafluorophosphate (Fluka, ≥ 97%) and diethyl sulfate (Fluka, 99%) were used without further purification. Dodecane (≥98%) and 1-decanol (≥98%) were obtained from Aldrich and used as received. Triethylene glycol monomethyl ether (purum, ≥ 97%) and sodium (in kerosene, purum) were obtained from Fluka and used without further purification. Chloroacetic acid was purchased from Riedel-de Haën (pure). Ortho-Phosphoric acid was applied as an 85% aqueous solution (Merck, p.a.). Sodium hydrogen carbonate (Merck, p.a.), potassium hydrogen carbonate (Roth, 99.5%) and lithium hydroxide (Aldrich, 98%) were used as received.

Solvents were of analytical grade, as obtained from Baker (ethanol), Merck (acetone, toluene) and Acros (dichloromethane), respectively. Dichloromethane was distilled from P_2O_5. Acetonitrile (Merck, gradient grade) was distilled from calcium hydride. Purified water was taken from a millipore system.

For SANS measurements, dodecane-d_{26} was obtained from Euriso-Top (98 %).

2. Synthesis

2.1. Ethylammonium nitrate (EAN)

Ethylammonium nitrate (EAN) was prepared by the reaction of equimolar amounts of ethylamine with nitric acid as described by Evans *et al.*[22] Water was first removed through rotary evaporation, followed by lyophilization. The obtained crude EAN was recrystallized trice from acetonitrile. The resulting colorless liquid was dried *in vacuo* ($p < 10^{-8}$ bar) at 50°C for one week and then stored under nitrogen atmosphere. Any occurrences of nitrous oxide impurities which, if present, would produce yellow discolorations,[192] could not be discerned.

III.2. Synthesis

The water content determined by coulometric Karl-Fischer titration of the final product was determined to be less than 100 ppm (m/m).

^{1}H-NMR (300 MHz, CD$_3$CN, TMS): δ_H: 7.35 (s, 3 H, NH$_3^+$), 3.03 (quart, 2 H, J = 7.41 Hz, CH$_3$), 1.23 (t, 3 H, J = 7.38 Hz, CH$_2$)

^{13}C-NMR (300 MHz, CD$_3$CN, TMS): δ_C: 35.24 (CH$_2$), 11.56 (CH$_3$)

Water content (Karl-Fischer titration): 48 ppm (m/m)

Melting point (DSC-onset): T_m = 14°C

Decomposition temperature (TGA-onset): T_d = 252°C

2.2. 1-Butyl-3-methylimidazolium tetrafluoroborate ([bmim][BF$_4$])

1-Butyl-3-methylimidazolium tetrafluoroborate ([bmim][BF$_4$]) was prepared according to a procedure described by Holbrey et al.[193] 1-Butyl-3-methylimidazolium chloride ([bmim][Cl]) was synthesized via the reaction of 1-methylimidazole with 1-chlorobutane in acetonitrile, and subsequently purified by recrystallizing trice from acetonitrile. The postmetathesis product was obtained from an ion exchange between [bmim][Cl] and sodium tetrafluoroborate in distilled ice-cooled water. [bmim][BF$_4$] was subsequently extracted with dichloromethane and dried under vacuum. The resulting colorless liquid was dried in a high-vacuum chamber ($p < 10^{-8}$ bar) at 50°C for one week, leading to the final product containing less than 60 ppm (m/m) of water.

1-Butyl-3-methylimidazolium chloride ([bmim][Cl])

^{1}H-NMR (300 MHz, CD$_3$CN, TMS): δ_H: 10.62 (s, 1 H, NCHN), 7.59 (t, 1H, J = 1.65 Hz, NCHCHN), 7.46 (t, 1H, J = 1.65 Hz, NCHCHN), 4.27 (t, 2H, J = 7.41 Hz, NCH$_2$CH$_2$), 4.07 (s, 3H, NCH$_3$), 1.84 (quint, 2H, J = 7.41 Hz, NCH$_2$CH$_2$CH$_2$), 1.32 (sext, 2H, J = 7.41, NCH$_2$CH$_2$CH$_2$CH$_3$), 0.90 (t, 3H, J = 7.40, CH$_2$CH$_3$)

^{13}C-NMR (300 MHz, CDCl$_3$, TMS): δ_C: 136.9, 123.1, 121.8, 48.7, 35.3, 31.3, 18.6, 12.3

ESI-MS: *m/z (+ESI):* m/z (+p): 138.9 (100%, C$^+$), 313.0 (1.14%, 2CA$^+$-Cluster)

Melting point (DSC-onset): T_m = 64°C

Water content (Karl-Fischer titration): 120 ppm (m/m)

1-Butyl-3-methylimidazolium tetrafluoroborate ([bmim][BF$_4$])

^{1}**H-NMR** (300 MHz, CD$_3$CN, TMS): δ_H: 8.52 (s, 1 H, NCHN), 7.43 (t, 1H, J = 1.78 Hz, NCHCHN), 7.38 (t, 1H, J = 1.8 Hz, NCHCHN), 4.14 (t, 2H, J = 7.31 Hz, NCH$_2$CH$_2$), 3.82 (s, 3H, NCH$_3$), 1.82 (quint, 2H, J = 7.40 Hz, NCH$_2$CH$_2$CH$_2$), 1.32 (sext, 2H, J = 7.43, NCH$_2$CH$_2$CH$_2$CH$_3$), 0.91 (t, 3H, J = 7.33, CH$_2$CH$_3$)

^{13}**C-NMR** (300 MHz, CD$_3$CN, TMS): δ_C: 135.7, 123.4, 122.0, 48.9, 35.5, 31.3, 18.6, 12.4

^{19}**F-NMR** (300 MHz, CD$_3$CN, TMS): -149.9 (s, ^{10}BF$_4^-$), -149.9 (s, ^{11}BF$_4^-$)

^{11}**B- NMR** (400MHz, CD$_3$CN, TMS): -0.50

ESI-MS: *m/z (+ESI):* m/z (+p): 138.9 (100.00%, C$^+$), 365.1 (1.02%, 2CA$^+$-Cluster)

Water content (Karl-Fischer titration): 55 ppm (m/m)

Decomposition temperature (TGA-onset): T_d = 391°C

2.3. 1-Butyl-3-methylimidazolium hexafluorophosphate ([bmim][PF$_6$])

The RTIL [bmim][PF$_6$] was obtained via metathesis of 1-butyl-methylimidazolium chloride ([bmim][Cl]) and potassium hexafluorophosphate, following a slightly modified procedure described by Cammarata et al.[194] [bmim][PF$_6$] was obtained from an ion exchange between [bmim][Cl] and potassium hexafluorophosphate in distilled ice-cooled water. [bmim][PF$_6$] was then extracted with dichloromethane and dried under vacuum. The resulting colorless liquid was dried under high-vacuum ($p < 10^{-8}$ bar) at 40°C for 5 days. Potentiometric titration of aqueous RTIL solution against a AgNO$_3$ standard solution yielded a halide mass fraction < $50 \cdot 10^{-6}$ for [bmim][PF$_6$].

^{1}**H-NMR** (300 MHz, CD$_3$CN, TMS): δ_H: 8.38 (s, 1 H, NCHN), 7.35 (m, 2H, NCHCHN), 4.11 (t, 2H, J = 7.11 Hz, NCH$_2$CH$_2$), 3.82 (s, 3H, NCH$_3$), 1.80 (quint, 2H, J = 7.38 Hz, NCH$_2$CH$_2$CH$_2$), 1.34 (sext, 2H, J = 7.68, NCH$_2$CH$_2$CH$_2$CH$_3$), 0.93 (t, 3H, J = 7.11, CH$_2$CH$_3$)

^{13}**C-NMR** (300 MHz, CD$_3$CN, TMS): δ_C: 135.6, 123.4, 122.0, 49.2, 49.0, 35.5, 31.3, 18.7, 12.3

^{19}**F-NMR** (300 MHz, CD$_3$CN, TMS): -72.7 (s), -70.20 (s)

Water content (Karl-Fischer titration): 60 ppm (m/m)

Decomposition temperature (TGA-onset): T_d = 437°C

III.2. Synthesis

2.4. 1-Alkyl-3-methylimidazolium chloride ([C$_n$mim][Cl], n = 12, 14, 16,18)

The surface active 1-alkyl-3-methylimidazolium chlorides [C$_n$mim][Cl] were obtained by a slightly modified procedure described in literature.[30,195] Equimolar amounts of 1-methylimidazole and 1-chloroalkanes (1-chlorodecane, 1-chlorotertadecane, 1-chlorohexadecane, 1-chlorooctadecane) were stirred between 5 to 14 days depending on the conversion in dry acetonitrile at 80°C under nitrogen atmosphere. The conversion was monitored via NMR. The raw products crystallized as slightly yellow solid and were recrystallized trice from tetrahydrofurane. The substances were subsequently dried under high vacuum ($p < 10^{-8}$ bar) at 30°C ([C$_{12}$mim][Cl]) and 40°C ([C$_{14\text{-}18}$mim][Cl]), respectively, for one week yielding white crystalline solids.

1-Dodecyl-3-methylimidazolium chloride ([C$_{12}$mim][Cl])

^{1}H-NMR (300 MHz, CDCl$_3$, TMS): δ_H: 10.54 (s, 1H, NCHN), 7.59 (t, 1H, J = 1.90 Hz, NCHCHN), 7.38 (t, 1H, J = 1.91 Hz, NCHCHN), 4.24 (t, 2H, J = 6.90 Hz, NCH$_2$CH$_2$), 4.11 (s, 3H, NCH$_3$), 1.84 (quint, 2H, J= 7.14 Hz, NCH$_2$CH$_2$CH$_2$), 1.27 (m, 20H, (CH$_2$)$_{10}$), 0.80 (t, 3H, J = 6.84 Hz, CH$_2$CH$_3$)

^{13}C-NMR (300 MHz, D$_2$O, TMS): δ_C: 123.75, 122.10, 49.51, 35.84, 31.83, 29.78, 29.61, 29.57, 29.39, 29.30, 28.95, 26.00, 22.53, 13.79

ESI-MS: *m/z (+ESI):* m/z (+p): 251 (100%, C$^+$), 538 (1.80%, C$_2$A$^+$ Cluster)

Melting point (DSC-onset): T_m = 41°C

Decomposition temperature (TGA-onset): T_d = 273°C

1-Tetradecyl-3-methylimidazolium chloride ([C$_{14}$mim][Cl])

^{1}H-NMR (300 MHz, CDCl$_3$, TMS): δ_H: 9.89 (s, 1H, NCHN), 7.55 (t, 1H, J = 1.65 Hz, NCHCHN), 7.53 (t, 1H, J = 1.65 Hz, NCHCHN), 4.20 (t, 2H, J = 7.14 Hz, NCH$_2$CH$_2$), 3.90 (s, 3H, NCH$_3$), 1.85 (quint, 2H, J= 7.13 Hz, NCH$_2$CH$_2$CH$_2$), 1.27 (m, 20H, (CH$_2$)$_{10}$), 0.86 (t, 3H, J = 7.11 Hz, CH$_2$CH$_3$)

^{13}C-NMR (300 MHz, D$_2$O, TMS): δ_C: 123.81, 122.08, 49.49, 35.89, 31.96, 29.90, 29.82, 29.61, 29.47, 29.15, 26.16, 22.63, 13.85

ESI-MS: *m/z (+ESI):* m/z (+p): 279 (100%, C$^+$), 593 (4.17%, C$_2$A$^+$ Cluster)

Melting point (DSC-onset): $T_m = 53°C$

Decomposition temperature (TGA-onset): $T_d = 280°C$

1-Hexadecyl-3-methylimidazolium chloride ([C$_{16}$mim][Cl])

^{1}H-NMR (300 MHz, CDCl$_3$, TMS): δ_H: 10.79 (s, 1H, NC**H**N), 7.43 (t, 1H, J = 1.92 Hz, NC**H**CHN), 7.28 (t, 1H, J = 1.92 Hz, NCHC**H**N), 4.29 (t, 2H, J = 7.41 Hz, NC**H$_2$**CH$_2$), 4.11 (s, 3H, NC**H$_3$**), 1.88 (quint, 2H, J = 7.14 Hz, NCH$_2$C**H$_2$**CH$_2$), 1.27 (m, 28H, (C**H$_2$**)$_{14}$), 0.86 (t, 3H, J = 7.14 Hz, CH$_2$C**H$_3$**).

^{13}C-NMR (300 MHz, D$_2$O, TMS): δ_C: 123.82, 122.07, 49.48, 35.91, 32.02, 30.05, 29.95, 29.71, 29.56, 29.23, 26.22, 22.67, 13.86

ESI-MS: *m/z (+ESI):* m/z (+p): 307 (100%, C$^+$), 593 (1.18%, C$_2$A$^+$ Cluster)

Melting point (DSC-onset): $T_m = 64°C$

Decomposition temperature (TGA-onset): $T_d = 274°C$

1-Octadecyl-3-methylimidazolium chloride ([C$_{18}$mim][Cl])

^{1}H-NMR (300 MHz, CDCl$_3$, TMS): δ_H: 10.82 (s, 1H, NC**H**N), 7.34 (t, 1H, J = 1.92 Hz, NC**H**CHN), 7.24 (t, 1H, J = 1.92 Hz, NCHC**H**N), 4.30 (t, 2H, J = 7.38 Hz, NC**H$_2$**CH$_2$), 4.11 (s, 3H, NC**H$_3$**), 1.89 (quint, 2H, J = 7.41 Hz, NCH$_2$C**H$_2$**CH$_2$), 1.27 (m, 30H, (C**H$_2$**)$_{16}$), 0.86 (t, 3H, J = 6.68 Hz, CH$_2$C**H$_3$**).

^{13}C-NMR (300 MHz, D$_2$O, TMS): δ_C: 123.82, 122.06, 49.47, 35.93, 32.05, 30.16, 29.95, 29.76, 29.56, 29.25, 26.24, 22.69, 13.88

ESI-MS: *m/z (+ESI):* m/z (+p): 335 (100%, C$^+$), 593 (1.19%, C$_2$A$^+$ Cluster)

Melting point (DSC-onset): $T_m = 70°C$

Decomposition temperature (TGA-onset): $T_d = 279°C$

2.5. 1-Ethyl-3-methylimidazolium ethylsulfate ([emim][EtSO$_4$])

1-Ethyl-3-methylimidazolium ethylsulfate ([emim][etSO$_4$]) was prepared in analogy to a protocol reported in literature.[196] First, diethylsulfate was added to 1-methylimidazole in toluene and the mixture was cooled in an ice bath under a nitrogen atmosphere. The reaction mixture was stirred overnight at room temperature, followed by decantation of the upper

III.2. Synthesis

organic phase. The lower IL phase was washed several times with toluene. Drying in high vacuum yielded the desired ionic liquid exhibiting a water content of less than 50 ppm.

¹H-NMR (300 MHz, CDCl$_3$, TMS): δ_H: 9.39 (s, 1H, NCHN), 7.49 (d, 2H, J = 1.62 Hz, NCHCHN), 4.24 (q, 2H, J = 7.38 Hz, NCH$_2$CH$_3$), 4.02 (q, 2H, J = 7.11, CH$_3$CH$_2$SO$_4^-$), 3.94 (s, 3H, NCH$_3$), 1.48 (t, 3H, J = 7.41 Hz, CH$_3$CH$_2$SO$_4^-$), 1.21 (t, 3H, J = 7.14, NCH$_2$CH$_3$)

¹³C-NMR (300 MHz, D$_2$O, TMS): δ_C: 139.13, 123.76, 121.96, 63.31, 45.12, 36.34, 15.48, 15.23

***ESI-MS**: m/z (+ESI)*: m/z (+p): 110.9 (57.14%, C$^+$), 181.0 (31.55%, 2CA$^+$-Cluster), 347.1 (100.00%, C$_2^+$C$^-$)

Water content (Karl-Fischer titration): 25 ppm (m/m)

Glass transition temperature: T_g = -80°C

Decomposition temperature (TGA-onset): T_d = 356°C

2.6. 2,5,8,11-Tetraoxatridecan-13-oic acid (TOTOA)

2,5,8,11-Tetraoxatridecan-13-oic acid (TOTOA) was prepared according to a modified procedure described by Matsushima et al.,[197] in high purity (> 99%, GC).

Sodium (30.44 g, 1.32 mol) was stepwise added to 360 mL of triethylene glycol monomethyl ether (TEGME) under nitrogen atmosphere. Dissolution of the sodium was achieved by vigorous stirring and gradual heating to 120°C. The formed hydrogen was casually removed by applying a slight N$_2$ flow. The clear yellowish solution was subsequently cooled to 100°C, and chloroacetic acid (62.91 g, 0.67 mol) dissolved in 110 mL TEGME was added dropwise within 20 minutes. Then, the reaction mixture was stirred at 100°C for 12 hours. After removing excess TEGME by distillation *in vacuo*, a brown suspension remained. Subsequent treatment with an aqueous solution of phosphoric acid (95.69 g, 0.98 mol) yielded a clear brown solution, to which 300 mL dichloromethane were added. The organic phase was separated, and the aqueous phase was extracted repeatedly with 100 mL dichloromethane. The unified organic phases were dried over magnesium sulfate. Filtration and solvent evaporation resulted in a slightly yellow liquid. The crude 2,5,8,11-tetraoxatridecan-13-oic acid was purified by threefold distillation (b.p. 135-145°C at 10^{-7} mbar), yielding 120.91 g of a clear colourless viscous liquid (81.2%).

¹H-NMR (300 MHz, CDCl$_3$, TMS): δ_H: 9.82 (s, 1 H, COOH), 4.08 (s, 2 H, CH$_2$COOH), 3.56

(m, 12 H, CH$_2$), 3.29 (s, 3 H, CH$_3$)

^{13}C-NMR (300 MHz, CDCl$_3$, TMS): δ$_C$: 173.00 (COOH), 71.89 (CH$_3$OCH$_2$), 71.20 (CH$_2$OCH$_2$COOH), 70.29 – 70.46 (4 C, CH$_2$), 68.74 (CH$_2$COOH), 58.76 (CH$_3$)

Elemental analysis: calculated: 48.64% C, 8.16% H; found: 47.56% C, 8.39% H

ESI-MS (H$_2$O / MeOH, NH$_4$Ac): m/z (+p): 102.9 (21 %), 176.8 (12 %, EH$^+$- HCOOH), 222.9 (57 %, EH$^+$), 240.0 (100 %, ENH$_4^+$); m/z (-p): 221.0 (100 %, E – H$^+$), 443.2 (10 %, 2E – H$^+$)

GC analysis: purity 99.3%

Water content (Karl-Fischer titration): 110 ppm (m/m)

2.7. TOTOA alkali salts

Alkali salts of 2,5,8,11-tetraoxatridecan-13-oic acid were prepared by direct neutralisation of the acid with alkali base. In the case of sodium and potassium, equimolar amounts of the corresponding hydrogen carbonate and the acid were dissolved in water and stirred for 1 hour. Lyophilization and subsequent drying *in vacuo* gave the desired salts in quantitative yields. [Na][TOTO] was obtained as a faintly yellow viscous liquid, [K][TOTO] as white crystals. The synthesis of the Li salt was carried out in 5:1 (v:v) mixture of ethanol and water using lithium hydroxide as base. After conversion, solvents were removed by lyophilization and the product was vacuum-dried, resulting in a highly viscous colourless liquid.

Lithium 2,5,8,11-tetraoxatridecan-13-oate ([Li][TOTO])

^{1}H-NMR (300 MHz, CDCl$_3$, TMS): δ$_H$: 3.90 (s, 2H, CH$_2$COOH), 3.62 (m, 10 H, CH$_2$) 3.53 (m, 2 H, CH$_2$), 3.35 (s, 3 H, CH$_3$)

^{13}C-NMR (400 MHz, CDCl$_3$, TMS): δ$_C$: 176.18 (COOH), 71.86 (CH$_3$OCH$_2$), 70.98 (CH$_2$COOH), 69.88 – 70.40 (4 C, CH$_2$), 69.33 (CH$_2$OCH$_2$COOH), 58.98 (CH$_3$)

Elemental analysis: calculated: 47.38% C, 7.51% H; found: 45.89% C, 7.87% H

ESI-MS (DCM-MeOH, NH$_4$Ac): m/z (+p): 222.9 (20%, EH$^+$), 228.9 (78%, MH$^+$), 236.0 (18%, MLi$^+$), 240.0 (19%, ENH$_4^+$), 267.0 (11%), 463.2 (100%, M$_2$Li$^+$), 691.4 (55%, M$_3$Li$^+$); m/z (-p): 221.0 (100%, E – H$^+$), 287.0 (22%, MCH$_3$COO$^-$), 449.2 (47%, 2M – Li$^+$)

Water content (Karl-Fischer titration): 103 ppm (m/m)

Glass transition temperature: T_g = -53°C

III.2. Synthesis

Decomposition temperature (TGA-onset): $T_d = 357°C$

Sodium 2,5,8,11-tetraoxatridecan-13-oate ([Na][TOTO])

^{1}H-NMR (300 MHz, CDCl$_3$, TMS): δ_H: 3.78 (s, 2 H, **CH$_2$**COOH), 3.58 (m, 10 H, CH$_2$), 3.48 (m, 2 H, CH$_2$), 3.30 (s, 3 H, CH$_3$)

^{13}C-NMR (300 MHz, CDCl$_3$, TMS): δ_C: 175.80 (COOH), 71.68 (CH$_3$O**C**H$_2$), 71.19 (**C**H$_2$COOH), 69.86 – 70.09 (4 C, CH$_2$), 69.12 (**C**H$_2$OCH$_2$COOH), 58.86 (CH$_3$)

Elemental analysis: calculated: 44.26% C, 7.02% H; found: 43.44% C, 6.46% H

ESI-MS (DCM-MeOH, NH$_4$Ac): m/z (+p): 222.9 (10%, EH$^+$), 244.9 (100%, MH$^+$), 266.9 (75%, MNa$^+$), 391.1 (22%), 511.2 (44%, M$_2$Na$^+$), 755.3 (19%, M$_3$Na$^+$), 849.3 (14%); m/z (-p): 221.0 (100%, E – H$^+$), 303.0 (31%, MCH$_3$COO$^-$), 465.2 (45%, 2M – Na$^+$), 709.4 (10%, 3M – Na$^+$)

Water content (Karl-Fischer titration): 211 ppm (m/m)

Glass transition temperature: $T_g = -57°C$

Decomposition temperature (TGA-onset): $T_d = 384°C$

Potassium 2,5,8,11-tetraoxatridecan-13-oate ([K][TOTO])

^{1}H-NMR (300 MHz, CDCl$_3$, TMS): δ_H: 3.78 (s, **CH$_2$**COOH), 3.56 (m, 5 H, CH$_2$), 3.48 (m, 2 H, CH$_2$), 3.31 (s, 3 H, CH$_3$)

^{13}C-NMR (300 MHz, CDCl$_3$, TMS): δ_C: 174.87 (COOH), 71.49 (CH$_3$O**C**H$_2$), 71.43 (**C**H$_2$COOH), 69.80 – 70.20 (4 C, CH$_2$), 69.00 (**C**H$_2$OCH$_2$COOH), 58.83 (CH$_3$)

Elemental analysis: calculated: 41.52% C, 6.58% H; found: 40.05% C, 6.87% H

ESI-MS (DCM-MeOH, NH$_4$Ac): m/z (+p): 223.0 (13%, EH$^+$), 240.1 (28%, ENH$_4^+$), 261.0 (100%, MH$^+$), 299.0 (61%, MK$^+$), 407.2 (17%), 521.2 (65%, M$_2$H$^+$), 559.2 (76%, M$_2$K$^+$), 781.3 (13%, M$_3$H$^+$), 819.3 (38%, M$_3$K$^+$); m/z (-p): 119 (44%), 144.9 (26%), 156.9 (10%), 221.0 (100%, E – H$^+$), 281.1 (39%, ECH$_3$COO$^-$), 319.0 (39%, MCH$_3$COO$^-$), 417.1 (13%), 443.2 (51%, 2E – H$^+$), 481.2 (44%, ME – H$^+$), 579.2 (15%, M$_2$CH$_3$COO$^-$), 741.4 (26%)

Water content (Karl-Fischer titration): 1292 ppm (m/m)

Melting point (DSC-onset): $T_m = 60°C$

Decomposition temperature (TGA-onset): T_d = 369°C

3. Experimental methods

3.1. Analytical methods

^{1}H-, ^{13}C-, ^{11}B- and ^{19}F-NMR spectra were recorded at 300 MHz or 400 MHz in CDCl$_3$, CD$_3$CN or D$_2$O using Bruker Avance 300 and Avance 400 spectrometers, respectively.

ESI-MS data were acquired on a ThermoQuest Finnigan TSQ 7000 spectrometer in dichloromethane-methanol (DCM-MeOH) mixtures to which a 10 mmol L^{-1} ammonium acetate (NH$_4$Ac) solution was added. For the imidazolium ILs the free cations are refered as C+, the anions as A$^-$. Throughout the analyses of mass spectrometry data, we refer to the free TOTOA acid as E, while the alkali salts are denoted as M.

Elemental analysis was performed by the Central Analytics Department at the University of Regensburg.

The water content of all products was determined by coulometric Karl-Fischer titration, using an Abimed MCI analyser (Model CA-02).

Differential scanning calorimetry (DSC) measurements of the 1-alkyl-3-methylimidazolium chloride ionic liquids were performed at a Perkin-Elmer DSC 7 with heating rate of 10 °C min^{-1}. The investigated temperature ranged from 25°C to 150°C. The DSC measurements of EAN were carried out on a Setaram microDSC III$^+$ within a temperature range between 10°C and 50°C applying a heating rate of 0.3°C min^{-1}. For the TOTO alkali salts DSC data were recorded operating a Mettler DSC 30 in a nitrogen atmosphere using Al crucibles. The [Li][TOTO] and [Na][TOTO] salts were investigated within a temperature range of (-150 - 20)°C, while the [K][TOTO] salt was measured from (-100 to 100)°C. The heating rate was in all cases 10 °C min^{-1}. Transition temperatures were generally obtained from heating curves, melting points were determined by onset analysis. Glass transition points were determined from the thermograms using the half-step temperature of the transition.

Infrared spectroscopy was performed on a Jasco FTIR 610 spectrometer using a attenuated total reflection (ATR) ZnSe crystal.

Thermal stability was studied using a thermogravimetric analyser from Perkin-Elmer TGA 7. Samples were measured at a heating rate of 10°C min^{-1}, applying a continuous nitrogen flow. Decomposition temperatures were determined using onset points of mass loss, being defined as the intersection of the baseline before decomposition and the tangent to the mass loss versus temperature.

The electrochemical stability of the salts was investigated by means of cyclic voltammetry (CV) measurements, employing platinum working and counter electrodes and Ag/Ag$^+$ (BAS) with Kryptofix® 22 (Merck, for synthesis) as reference electrode. For all measurements, samples were dissolved in dry acetonitrile (Merck, for DNA synthesis, ≤ 10 ppm H$_2$O). Scans were recorded under inert gas starting in anodic direction, with a rate of 10 mV s^{-1}. Sincere thanks are given to Dr. C. Schreiner who performed the CV measurements of the TOTO alkali salts.

Gas chromatography (GC) was performed on a HP 6890 analyser equipped with an autosampler and an FID detector. A HP-5 column type was used for analysing TOTOA batches, the carrier gas was high-purity helium. Samples were injected at 250°C, peaks were detected at 300°C. Dipl. Chem. B. Ramsauer is gratefully acknowledged for help with the GC measurements.

3.2. Electrical conductivity

Conductivity measurements of the pure ionic liquids have been performed within a temperature range between -25°C and 195°C in intervals of 10°C. Conductivity measurements of the high temperature stable microemulsions have been performed between 30°C and 150°C in intervals of 30°C.

Measurements of the pure ILs in the low temperature region (-25 - 30)°C were carried out with an in-house built apparatus described by Barthel *et al.* equipped with a precision thermostat, symmetrical Wheatstone bridge with Wagner earth, sine generator and resistance decade.[198,199] A similar apparatus constructed for conductivity measurements up to 150°C was used for the high temperature region of the microemulsions (60 - 150)°C. For the conductivity measurements of pure ILs at high temperatures (35- 195)°C a precision thermostat combined with symmetrical Wheatstone bridge with Wagner earth, sine generator and resistance decade was set up. Both thermostats were stable to < 0.003°C over the investigated temperature range, a set of six three-electrode capillary (set1) cells with cell constants ranging from 2 m^{-1} to 1161 m^{-1} and a set of four two electrode capillary cells (set2) with cell constants, B, ranging from 1260 m^{-1} to 3400 m^{-1} were used. Cell constants were determined with aqueous KCl solutions according to a procedure described by Wachter *et al.*[200] Resistances were measured at frequencies between 480 Hz and 10 kHz from which the value R at infinite frequency was

III.3. Experimental methods

obtained by linear extrapolation.[200] Repeated measurements of selected samples agreed within 0.3 %. The relative uncertainty of the electrical conductivities was estimated to be less than 0.5 %. The obtained cell constants are summarized in Table III-1.

Table III-1. Cell constants at 25°C obtained from calibration with aqueous KCl solutions.

Cell number	Set1 B / m^{-1}	Set 2 B / m^{-1}
1	2.084	1260
2	24.61	2440
3	53.6	4613
4	224.0	3400
5	470	-
6	1161	-

The temperature dependence of cell constants is affected by the expansion coefficient of platinum and Pyrex glass and can be expressed as

$$B(T) = B(298.15\,K)(1+\beta(T-298.15\,K)) \quad \text{(III-1)}$$

with the temperature coefficient which can be written as:

$$\beta = \frac{1}{B(298.15K)}\left(\frac{dB}{dT}\right) \quad \text{(III-2)}$$

The temperature coefficient is therefore dependent on the type of cell used. It was first described by Robinson and Stokes[201] and experimentally verified by Barthel and coworkers.[198,200] For capillary cells β was found to be $-3.5 \cdot 10^{-6}$ K^{-1}. As the application of β changed the electrical conductivities only within the given uncertainty limits, the temperature dependence of the cell constants was not taken into account.

3.3. Dynamic light scattering

Dynamic light scattering (DLS) experiments were carried out at a CGS-II goniometer from ALV (Germany). All measurements were performed at a scattering angle of 90°. A 22 mW He-Ne laser working at 632.8 nm and stabilized by a voltage regulator served as light source. The measurement temperature was adjusted with a Lauda RS 6 thermostat, the sample temperature was controlled with two platinum resistance thermometers that were placed in the toluene bath surrounding the sample. The accuracy in temperature was better than ± 0.1°C. For each measurement standard glass cuvettes, cleaned with a cuvette cleaning apparatus in order to avoid any impurities or dust in the sample, were used. Furthermore, the microemulsions were filtered directly into the cuvettes with cellulose acetate syringe filters (0.2 µm pore diameter) to avoid the presence of dust. The software simultaneously gave the temporal averaged intensities and the intensity time-correlation function from the input signals of the detector. For each measurement 10 runs each of 45 s duration have been performed, the mean value of the ten runs was taken for the data analysis. The filling level of the toluene bath surrounding the sample was controlled, before starting a new measurement series.

The accuracy of the measurements was controlled by measuring several polystyrene standards in water. Therefore, three different standards with diameters d = 100nm ± 3nm (Fluka), d = 220 nm ± 6nm (Duke Scientific Corp.) and d = 290 nm ± 8 nm (Interfacial Dynamic Corp.) were diluted with water until a slightly bluish solution was obtained. For each standard an intensity autorrelation function was obtained exhibiting a single exponential decay. These functions are shown in Figure III-1, full lines fit with a single exponential decay according to eq. II-29. The non-weighted size distributions of the standards are shown in Figure III-2.

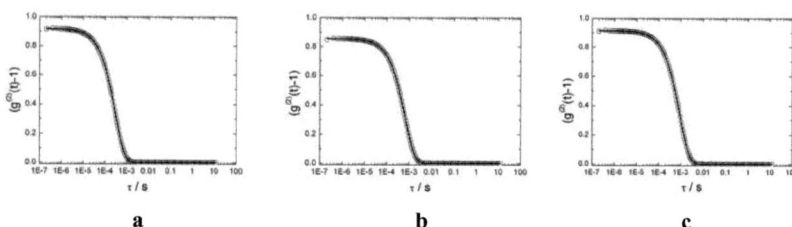

a b c

Figure III-1. Intensity autocorrelation functions for polystyrene standards with diameters, d = 100nm (a), d = 220 nm (b) and d = 290 nm (c), full lines fit with eq. II-29.

III.3. Experimental methods

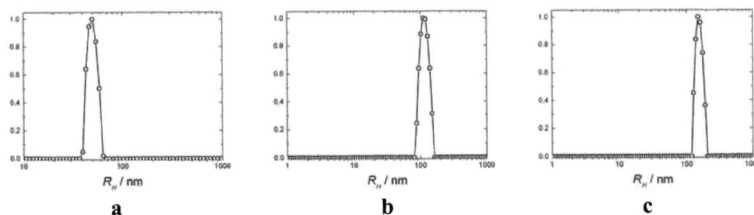

Figure III-2. Distribution functions for polystyrene standards $d = 100$ nm (a), $d = 220$ nm (b) and $d = 290$ nm (c).

The results for the radii from the exponential fit, the cumulant and the maxima of the distribution functions, R_H^{max}, as well as the PDI are summarized in Table III-2. All results are in good agreement with the sizes certified by the manufacturer within the given uncertainty limits confirming the accuracy of the DLS apparatus.

Table III-2. Results from the data analysis of the polystyrene standards at 25°C.

R_H(manufacturer) / nm	50	110	145
	Cumulant fit		
R_H(first order) / nm	50.8	114.0	154.2
R_H(second order) / nm	50.2	113.4	153.9
R_H(third order) / nm	50.3	114.0	155.0
PDI	0.033	0.016	0.005
	Single exponential fit		
R_H / nm	50.7	113.9	154.2
	Distribution		
R_H^{max} / nm	50.4	110.0	150.0

3.4. Densities

Densities, ρ, were determined using a vibrating tube densimeter (Anton Paar DMA 60) within temperatures ranging from 25°C to 65°C. The calibration of the instrument was performed for

III. Experimental

each temperature with purified dry nitrogen and degassed water with precise density values available from literature.[202] The uncertainty of ρ was estimated to be less than 0.1 kg m^{-3}.

For the high temperature stable microemulsions densities between 30°C and 150°C were required. Therefore, densities of the pure EAN, dodecane, biodiesel and the densities of the [C$_{16}$mim][Cl]+decanol mixture (1:4, molar ratio) were measured with a pyknometer at (25 - 150) °C in intervals of 10°C. The pyknometer was calibrated with degassed water at temperatures ranging from 20°C to 50°C in intervals of 10°C, no temperature effect on the pyknometer volume could be detected within its standard deviation.

3.5. Viscosities

Viscosity measurements were carried out on a Bohlin rheometer (type CVO 120 High Resolution) under argon atmosphere at controlled temperature. The temperature calibration was performed with a standard viscosity oil and a platinum resistance thermometer within a temperature range between (20- 65)°C. The accuracy in temperature was estimated to be better than ± 0.2°C. The microemulsions were measured at 30°C, [Na][TOTO] in the range of (20- 65)°C, working with a CP40/4° cone. Samples were studied at shear rates ranging from (0.25 to 200) s^{-1}, except for the measurement of [Na][TOTO] at 20°C, which had to be stopped at 50 s^{-1} due to the high viscosity.

3.6. Small angle X-Ray scattering (SAXS)

SAXS experiments for microemulsions with dodecane at 30°C were performed in flat cells of 0.1 mm and 0.2 mm thicknesses with Kapton windows on the Huxley-Holms High Flux camera.[203,204] The X-ray source is a copper rotating anode operating at 15 kW. The K_α radiation (λ = 0.154 nm) is selected by a Xenocs monochromator mirror. Spectra were recorded with a two-dimensional gas detector of 0.3 m in diameter giving an effective q-range of (0.2 to 3.5) nm^{-1} ($q = 4\pi / \lambda \, sin(\theta / 2)$), where θ is the scattering angle and λ the wavelength of the incident beam. Data correction, radial averaging, and absolute scaling were performed using routine procedures.[205]

Small angle X-ray scattering data for microemulsions with biodiesel were obtained using a XENOCS set-up. The X-ray beam originates from a Mo GENIX source. The K_α radiation (λ=0.71 Å^{-1}) is selected using a multilayered curved mirror (one reflexion) focusing the beam at the infinite. The size of the beam (less than 1mm^2) in front of the sample is defined by

III.3. Experimental methods

scatterless slits provided by FORVIS.[206] Sample and empty cell transmissions are determined using an offline pin-diode that can be inserted downstream the sample. The sample distance to the 2D MAR345 is 750 cm. Small angle scattering experiments covered a q-range between (0.02 and 4.0) nm^{-1}. Quartz capillaries are used as sample containers; usual corrections for background (empty cell and detector noise) subtractions and intensity normalization using Lupolen^{TM} as standard were applied.

3.7. Small angle neutron scattering (SANS)

SANS experiments were carried out on the instrument D22 at the Institut Laue-Langevin. Quartz cells from Hellma were used, which are stable up to 8 bars at 300°C with a height of 50 mm, a length of 40 mm and a thickness of 1 mm. SANS spectra were recorded at 30°C, 60°C, 90°C and 150°C. Samples were thermostated in a cupper block allowing rapid heating and cooling of the samples. Sample temperatures were measured with a platinum resistance thermometer from a reference cell, the temperature accuracy was estimated to be better than ± 0.5°C. Three configurations were used to cover a q-range between (0.07 and 4.0) nm^{-1}. Experiments were carried out at a wavelength of 0.6 nm with a relative wavelength spread $\Delta\lambda/\lambda$ of 10%.[207]

A schematic representation of the steady-state instrument D22 at the Institut Laue Langevin (ILL) is given in Figure III-3.

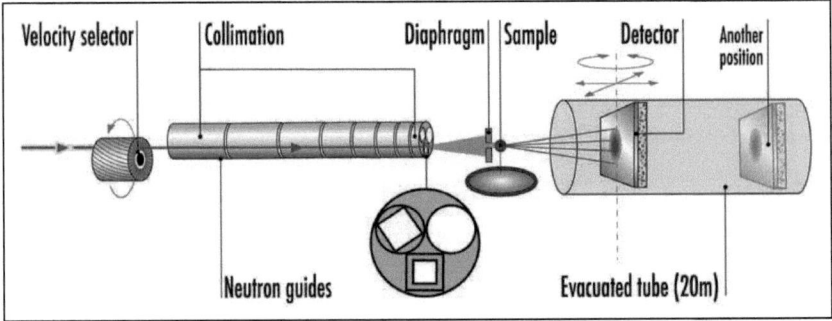

Figure III-3. Schematic representation of the D22 instrument, ILL, Grenoble, France.[208]

By the horizontal cold source of the reactor, a white beam is produced. The maximum neutron flux at the sample is $1.2 \cdot 10^8$ cm^{-2} s^{-1}. The wavelength can be adjusted by a mechanical velocity selector consisting of a rotating drum with helically curved absorbing slits at its surface. The rotation speed can be varied between 28000 to 4000 runs per minute, resulting in wavelengths between (0.46 and 4.0) nm. The wavelength calibration is performed with silver behenate as it exhibits narrow Bragg peaks. The collimation part on D22 is composed by eight guides having a cross section of (55·40) mm². D22 possesses the largest area multi-detector (^{3}He) of all small-angle scattering instruments (active area (96·96) cm^2 with pixel size (7.5·7.5) mm^2).[209] It moves inside a 2.5 m wide and 20 m long vacuum tube providing sample detector distances of (1.35 to 18) m. It can be translated laterally by 50 cm, and rotated around its vertical axis to reduce parallax. The beam stop preventing detector damages are made of B$_4$C and Cd with flexible sizes depending on the sample geometry.[209]

The raw data obtained from the SANS measurements were radial averaged, corrected for electronic background and empty cell using standard ILL software. Particular care was taken for the normalization using water scattering.

The smearing of the scattered intensity can be subdivided in three factors, the infinite size of the incident beam, the wavelength resolution and the detector pixel size.[210] The pixel size has a negligible effect. The q resolution can be expressed as

$$\Delta q = -q\left(\frac{\delta\lambda}{\lambda}\right) + \left(\frac{4\pi}{\lambda}\right)\cos\theta\Delta\theta \qquad \text{(III-3)}$$

where λ is the wavelength of the incident beam and θ the scattering angle. Hence, Δq^2 can be written as

$$\begin{aligned}\Delta q^2 &= q^2\left(\frac{\delta\lambda}{\lambda}\right)^2 + \left(\frac{4\pi}{\lambda}\right)^2\cos^2\theta\Delta\theta^2 = \Delta q^2(\lambda) + \Delta q^2(\theta) \\ &= q^2\left[\left(\frac{1}{2\sqrt{2\ln 2}}\frac{\Delta\lambda}{\lambda}\right)^2\right] + \left[\left(\frac{4\pi}{\lambda}\right)^2 - q^2\right]\Delta\theta^2\end{aligned} \qquad \text{(III-4)}$$

Where $\Delta\lambda/\lambda$ is related to the full width at half maximum (FWHM) of the triangular function describing the wavelength distribution by FWHM = $\lambda_0(\Delta\lambda/\lambda)$ and $\Delta\theta$ to the width of the incident beam. A triangular function centered on λ_0, the wavelength of the incident beam can be expressed as:

III.3. Experimental methods

$$T(\lambda) = \frac{1}{[\lambda_0 (\Delta\lambda/\lambda)]^2}\lambda + \frac{\lambda_0 (1-(\Delta\lambda/\lambda))}{[\lambda_0 (\Delta\lambda/\lambda)]^2} \quad \lambda \leq \lambda_0 \qquad \text{(III-5)}$$

$$T(\lambda) = \frac{-1}{[\lambda_0 (\Delta\lambda/\lambda)]^2}\lambda + \frac{\lambda_0 (1+(\Delta\lambda/\lambda))}{[\lambda_0 (\Delta\lambda/\lambda)]^2} \quad \lambda \geq \lambda_0 \qquad \text{(III-6)}$$

For a wavelength of λ_0 = 0.6 nm the wavelength distribution of 10% in terms of a triangular function is illustrated in Figure III-4.

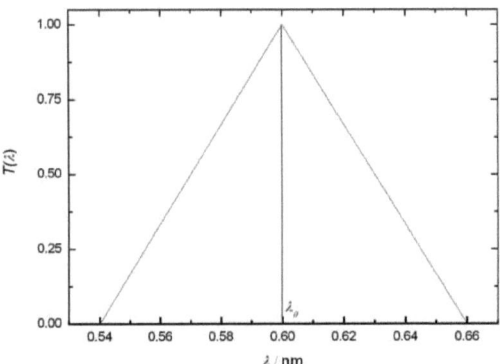

Figure III-4. Triangular function used to describe the wavelength distribution with λ_0 = 0.6 nm and $\Delta\lambda/\lambda$ = 10 %

IV. Results and Discussion

1. The conductivity of imidazolium-based ionic liquids over a wide temperature range. Variation of the anion

The application of room temperature ionic liquids (RTILs) in high temperature stable colloidal systems such as microemulsions requires knowledge of the physicochemical properties of the pure compounds. One important method used to characterize the high temperature stable microemulsions was temperature dependent conductivity. Therefore, prior knowledge of important transport properties, such as conductivity of the ILs is essential. Furthermore, the purity of the substances and the accuracy of the data can be evaluated by comparing the results to available literature data.

1.1. Abstract

The effect of counter ion on the electrical conductivity, κ, of 1-butyl-3-methylimidazolium ([bmim]) based room temperature ionic liquids (RTILs) was studied over a wide temperature range. Data for κ of [bmim][DCA], [bmim][PF$_6$], [bmim][TA] and [bmim][TfO] over a maximum temperature range of (248 to 468) K, depending on the limitations caused by the crystallization temperature and the thermal stability, respectively, are reported. The uncertainty in κ is estimated to be less than 0.5 %. The obtained conductivity values are well described by the Vogel-Fulcher-Tammann equation. To ensure the high thermal stability, degradation temperatures of the RTILs were determined via thermogravimetric analyses. Additionally, densities, ρ, and the corresponding molar conductivities, Λ, are reported for the temperature range of (278 to 338) K. Deviations from the ideal KCl line were illustrated in terms of a Walden plot that indicate that the investigated salts can be classified as high-ionicity ionic liquids.

1.2. Introduction

Over the last years Ionic Liquids (ILs) have gained more and more attention, because of their unique properties such as negligible vapor pressures, high electrical conductivities and thermal stability, coupled with a wide liquid range.[17,16] These properties render ILs, especially room temperature Ionic Liquids (RTILs), excellent candidates for potential applications.[3,4] Ionic liquids can be applied as solvents in organic reactions or catalysis[7,8] and as media in

IV.1. Conductivity of room temperature ionic liquids over a wide temperature range

extraction processes.[11] Other studies concern the use of ionic liquids as electrolytes in batteries,[12] double layer capacitors or solar cells.[13,14]

In recent years, progress has been made in the determination of thermodynamic and transport properties of RTILs, with imidazolium-based compounds being in the focus of many studies. Various groups studied the conductivity, κ, and viscosity, η, of imidazolium based RTILs.[40,211-218] Yoshida et al. described κ and η of different RTILs based on the dicyanamide anion at 293 K,[215] whereby Hunger et al. documented electrical conductivity of 1-butyl-3-methylimidazolium dicyanamide ([bmim][DCA]) over a temperature range of (278 to 338) K.[219] The electrical conductivity of 1-butyl-3-methylimidazolium hexafluorophosphate ([bmim][PF$_6$]) has been reported by Widegren et al. (288 – 323) K,[213] Tokuda et al. (263 – 373) K[214] and Suarez et al. (273 – 293) K.[211] Moreover, κ of 1-butyl-3-methylimidazolium trifluoroacetate ([bmim][TA]) has been described at 293 K[40] and at temperatures ranging from (263 to 373) K.[214] Electrical conductivity data of 1-butyl-3-methylimidazolium trifluoromethanesulfonate ([bmim][TfO]) have been determined by Tokuda et al. between 263 K and 373 K[214] and by Bonhôte et al. at 293 K.[40]

Nevertheless, to the best of our knowledge, there is no study that benefits from the wide liquid range and high thermal stability of RTILs and describes ionic conductivities over a wide temperature range. The physicochemical properties of RTILs, such as conductivity, depend on both, the nature and size of cation and anion. Herein, we study the effect of anion on the conductivity of RTILs based on the [bmim] cation. Precise temperature dependent conductivity data of [bmim][DCA] (248 - 468) K, [bmim][PF$_6$] (268 - 468) K, [bmim][TfO] (268 K- 468 K), and [bmim][TA] (248 K- 368 K) are reported. This study complements an investigation of Stoppa et al.[263] on the cation-dependence of κ for a series of imidazolium tetrafluoroborates.

1.3. Synthesis and sample handling

The RTIL [bmim][PF$_6$] was synthesized as described in section III.2.3. The RTILs [bmim][DCA], [bmim][TfO] and [bmim][TA] were synthesized by A. Stoppa and J. Hunger who are gratefully acknowledged here.

[bmim][DCA] was obtained by the metathesis reaction of 1-butyl-3-methylimidazolium chloride ([bmim][Cl]) and sodium dicyanamide, following a slightly modified procedure described by Cammarata et al.[194] A detailed description of the synthesis is given elsewhere.[218]

The RTILs [bmim][TfO] and [bmim][TA] were prepared by slowly adding a slight molar excess (~1:1.1) of methyl trifluoromethanesulfonate or methyl trifluoroacetate, respectively, to a solution of 1-butylimidazole in acetonitrile. Excess reactants were removed by vacuum distillation.

All RTILs were dried under high vacuum ($p < 10^{-8}$ bar) at 40°C for one week yielding slightly yellow liquids with water mass fractions < 80·ppm (m/m). Potentiometric titration of freshly prepared aqueous RTIL solutions against standardized $AgNO_3$(aq) yielded halide mass fractions < 50·10^{-6} for [bmim][PF_6] and = 0.0021 for [bmim][DCA], respectively. No AgCl precipitation could be observed when adding $AgNO_3$(aq) to [bmim][TA] and [bmim][TfO]. No contaminations were detected with ^{1}H-, ^{13}C- and (where applicable) ^{19}F-NMR. The samples were stored in a N_2-filled glovebox; conductivity cells were filled under nitrogen atmosphere.

1.4. Results and discussion

The data for the electrical conductivities of the four RTILs are summarized in Table IV-1.

[bmim][DCA] was investigated over the full range of (248 to 468) K, whereas [bmim][PF_6] and [bmim][TfO] could only be studied between (268 and 468) K, as samples crystallized in the conductivity cells when cooling to 258 K. For [bmim][TA] data are reported between (248 and 368) K, since the formation of bubbles was observed above 375 K. This effect may be related to a partial degradation of [bmim][TA] at these temperatures, although TGA experiments yielded a decomposition temperature of T_d = 461 K. As already small amounts of impurities, including degradation products, can have significant effects on the conductivity of [bmim][TA], we abstain from reporting κ values for T > 368 K. The electrical conductivity decreases over the whole investigated temperature range in the order [bmim][DCA] > [bmim][PF_6] > [bmim][TfO] > [bmim][TA].

The present data for [bmim][DCA] (Figure IV-1) are in reasonable agreement with the κ value given by Yoshida et al. for 298 K.[215] However, they are systematically lower, with values for the relative deviation, $\delta_{Lit} = (\kappa - \kappa_{Lit}) / \kappa \cdot 100$, of $|\delta_{Lit}| \leq 11.3$, than those reported by Hunger et al. (Figure IV-1).[219] A possible reason is the amount of halide impurities, for which ref. 219 reports a mass fraction of 0.005 whereas that of the present sample is only 0.0021.

IV.1. Conductivity of room temperature ionic liquids over a wide temperature range

Table IV-1. Conductivities, κ, of the investigated RTILs.

T/K	$\kappa / \mathrm{S \cdot m^{-1}}$			
	[bmim][DCA]	[bmim][PF$_6$]	[bmim][TfO]	[bmim][TA]
248.15	0.0483			0.00858
258.15	0.1180			0.0244
268.15	0.242	0.0164	0.0539	0.0569
278.15	0.433	0.0385	0.1038	0.1144
288.15	0.702	0.0792	0.1808	0.205
298.15	1.052	0.1465	0.290	0.332
308.15	1.483	0.248	0.436	0.504
318.15	1.991	0.390	0.621	0.723
328.15	2.57	0.578	0.848	0.988
338.15	3.22	0.815	1.117	1.301
348.15	3.93	1.102	1.426	1.658
358.15	4.69	1.440	1.775	2.06
368.15	5.41	1.825	2.16	2.50
378.15	6.34	2.26	2.60	
388.15	7.26	2.74	3.04	
398.15	8.19	3.25	3.52	
408.15	9.15	3.81	4.03	
418.15	10.12	4.39	4.56	
428.15	11.12	5.01	5.12	
438.15	12.10	5.65	5.69	
448.15	13.09	6.31	6.28	
458.15	13.75	7.00	6.88	
468.15	14.54	7.69	7.50	

IV. Results and discussion

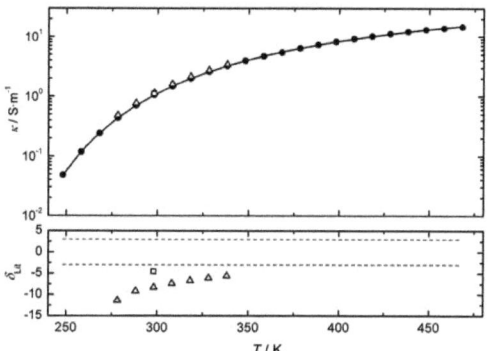

Figure IV-1. Specific conductivities of [bmim][DCA] from this study (●) and relative deviations, δ_{Lit}, of data reported by Hunger et al. (△[219]) and Yoshida et al. (□[215]) Full lines show a fit of κ with the VFT equation, the corresponding fit parameters are given in Table IV-2, dashed lines indicate a (arbitary) $\delta_{Lit} = \pm 3$ margin.

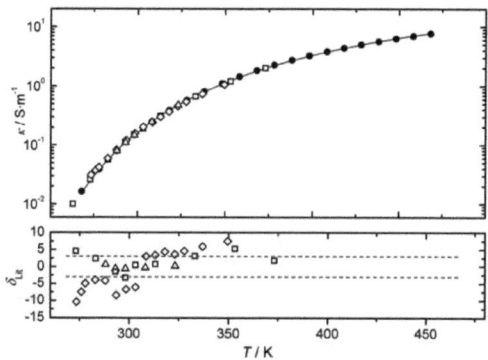

Figure IV-2. Specific conductivities of [bmim][PF$_6$] (●) compared to data documented by Widegren et al. (△[213]), Tokuda et al. (□[214]) and Suarez et al. (◊[211]) and relative deviations, δ_{Lit}, of interpolated values from the published data. Full lines show a fit of κ with the VFT equation, the corresponding fit parameters are given in Table IV-2, dashed lines indicate a (arbitary) $\delta_{Lit} = \pm 3$ margin.

For [bmim][PF$_6$] (Figure IV-2) conductivity data were reported by Widegren et al.[213] between (288.15 and 323.15) K and by Suarez et al.[211] between (273.50 and 349.50) K. Whereas the

IV.1. Conductivity of room temperature ionic liquids over a wide temperature range

former are in excellent agreement with our results (-0.6 ≤ δ_{Lit} ≤ 0.6) the latter deviate considerably (-10.4 ≤ δ_{Lit} ≤ 7.5). Conductivities published by Tokuda et al.[214] for the range (273.15 to 373.15 K) also agree well (-3.4 ≤ δ_{Lit} ≤ 5.2) except for the point at 263.15 K (δ_{Lit} = -144.9).

Except for the lowest temperature, conductivities reported by Tokuda et al.[214] for [bmim][TfO] (Figure IV-1) in the range (263.15 and 373.15) K are in very good agreement with the present data (|δ_{Lit}| ≤ 3.8). The value reported by Bonhôte et al.[40] for 293 K deviates again considerably (δ_{Lit} ≤ -60).

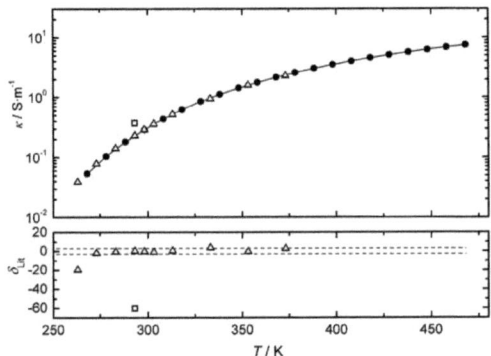

Figure IV-3. Specific conductivities of [bmim][TfO] (●) compared to data documented by Tokuda et al. (Δ[214]) and Bonhôte et al. (□[40]) and relative deviations, δ_{Lit}, of interpolated values from the published data. Full lines show a fit of κ with the VFT equation, the corresponding fit parameters are given in Table IV-2, dashed lines indicate a (arbitary) δ_{Lit} = ±3 margin.

Also for [bmim][TA] comparison of conductivity data is only possible with the results of Tokuda et al.[214] in the range (263.15 and 373.15) K and with Bonhôte et al.[40] for 293 K (Figure IV-4). Whilst the former are all systematically smaller (δ_{Lit} ≤ 10.4)[214] the latter is larger by 22%.[40]

To ensure the high thermal stability of the RTILs, the decomposition temperatures (T_d) have been determined via thermogravimetric analysis. The obtained values from onset analysis were T_d = 566 K for [bmim][DCA], T_d = 710 K for [bmim][PF$_6$], T_d = 689 K for [bmim][TfO] and $T_d = 461$ K for [bmim][TA]. These values are in quite good agreement compared to literature data $T_d = 573$ K[220] for [bmim][DCA], $T_d = 706$ K[212] for [bmim][PF$_6$], $T_d = 682$[212], 665 K[220] for [bmim][TfO] and $T_d = 449$ K[212] for [bmim][TA].

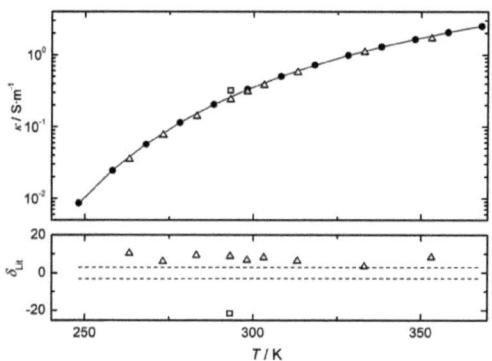

Figure IV-4. Specific conductivities of [bmim][TA] (●) compared to data documented by Tokuda et al. (Δ[214]) and Bonhôte et al. (□[40]) and relative deviations, δ_{Lit}, of interpolated values from the published data. Full lines show a fit of κ with the VFT equation, the corresponding fit parameters are given in Table IV-2, dashed lines indicate a (arbitary) $\delta_{Lit} = \pm 3$ margin.

The conductivity of RTILs, especially for those that show a glass transition temperature (T_g) can often be described by the empirical Vogel-Fulcher-Tammann eq. with the fit parameters A and B and T_0, where the latter is the so called VFT temperature.[52]

$$\ln (\kappa \,/\, \text{S m}^{-1}) = \ln (A \,/\, \text{S m}^{-1}) + B \,/\, (T - T_0) \qquad \text{(IV-1)}$$

The conductivity data for the four RTILs were found to be well described by the VFT equation. The values obtained for A, B and T_0, are summarized in Table IV-2, together with the corresponding uncertainty of the fit, σ. The conductivity data compared to literature data with the relative deviations as well as the corresponding VFT fit are illustrated for [bmim][DCA] (Figure IV-1), [bmim][PF$_6$] (Figure IV-2), [bmim][TfO] (Figure IV-3) and

IV.1. Conductivity of room temperature ionic liquids over a wide temperature range

[bmim][TA] (Figure IV-4).

Table IV-2. Parameters of the VFT fit with the standard uncertainty of the overall fit, σ.

	[bmim][DCA]	[bmim][PF$_6$]	[bmim][TfO]	[bmim][TA]
A / S·m^{-1}	104.3	131.1	95.1	103.5
B / K	-572.0	-818.1	-767.2	-737.3
T_0 / K	173.7	177.4	165.6	169.8
$100 \cdot \sigma$	1.0	1.6	0.3	0.7

Consistent with the literature[52] the VFT temperature (Table 2) is ~20-30 K below the glass-transition temperature, T_g, determined via differential scanning calorimetry, namely T_g = (179 K;[212] 183 K[220]) for [bmim][DCA]; (196 K; [212] 197 K;[220] 196 K[221]) for [bmim][PF$_6$] ; 191 K[212] for [bmim][TfO] and T_g = 195 K[212] for [bmim][TA]. Tokuda and coworkers reported values of T_0 = 174 K for [bmim][PF$_6$], 162 K for [bmim][TfO] and 172 K for [bmim][TA].[212] Keeping in mind the different temperature ranges of both studies the agreement of these T_0 with the VFT temperatures of Table IV-2 is very good and reflects the generally excellent agreement of the experimental conductivities of both investigations. In contrast to the behavior of the [C$_n$mim][BF$_4$] compounds studied in the companion study, where a linear dependence of the strength parameter B/T_0 on the number of carbon atoms, n, in the N-alkyl chain was observed,[263] no general trend in A, B, T_0, or B/T_0 can be found when varying the anion of [bmim]$^+$ RTILs.

To get more information about the transport properties of the RTILs densities, ρ, were measured between 278.15 K and 338.15 K, a viscosity correction as described by Heintz *et al.*[222] was not taken into account, since reliable viscosity data and/or more sophisticated correction procedures are currently not available, we refrain from giving corrected ρ values. The respective values and the corresponding molar conductivities, Λ_m, are summarized in Table IV-3. The relation of fluidity to conductance can be considered in terms of a Walden plot of the data, as described by Angell *et al.*[52,55,56] The Walden rule relates the ion mobilities to the fluidity of the medium: if a liquid substantially consists of independent ions only, then

IV. Results and discussion

the Walden plot will be close to an ideal line, which is represented by potassium chloride.[56] Substances whose plot lies more than one order of magnitude below the ideal line can in this context be classified as "poor" ionic liquids (compare Figure I-2). The Walden plots for the RTILs [bmim][PF$_6$], [bmim][TfO] and [bmim][TA], where fluidities, η^{-1}, are available are illustrated in Figure IV- 5. Viscosities were calculated from the viscosity VFT parameters reported by Tokuda et al.[212] Despite some anion-specific deviations, all data are close to the "ideal" KCl line representing complete dissociation.[54,222,221] Thus, the RTILs [bmim][PF$_6$], [bmim][TfO] and [bmim][TA] can be classified as high-ionicity ionic liquids.

Table IV-3. Densities, ρ, and molar conductivities, Λ_m, of the investigated RTILs.

T / K	ρ / kg·m^{-3}	$10^4 \cdot \Lambda_m$ / S·m^2·mol^{-1}	ρ / kg·m^{-3}	$10^4 \cdot \Lambda_m$ / S·m^2·mol^{-1}	ρ / kg·m^{-3}	$10^4 \cdot \Lambda_m$ / S·m^2·mol^{-1}	ρ / kg·m^{-3}	$10^4 \cdot \Lambda_m$ / S·m^2·mol^{-1}
	[bmim][DCA]		[bmim][PF$_6$]		[bmim][TfO]		[bmim][TA]	
278.15	1072.2	0.830	1383.4	0.1021	1313.2	0.228	1230.4	0.235
288.15	1066.1	1.352	1376.9	0.211	1305.5	0.399	1223.4	0.422
298.15	1059.6	2.038	1368.3	0.393	1297.4	0.644	1215.5	0.692
308.15	1053.4	2.89	1360.0	0.668	1289.3	0.974	1207.7	1.053
318.15	1047.3	3.90	1351.6	1.058	1281.6	1.398	1200.2	1.519
328.15	1040.9	5.07	1343.2	1.577	1273.6	1.920	1192.6	2.09
338.15	1034.7	6.39	1334.7	2.24	1265.8	2.54	1185.1	2.77

IV.1. Conductivity of room temperature ionic liquids over a wide temperature range

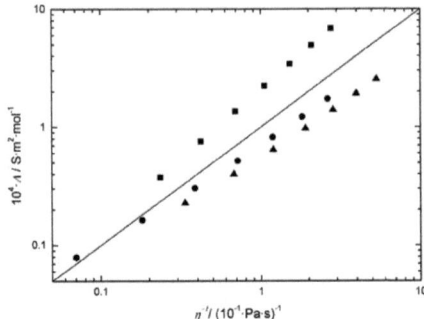

Figure IV- 5. Walden plot for [bmim][PF$_6$] (●), [bmim][TfO] (▲) and [bmim][TA] (■). Required viscosities were calculated from the VFT parameters given in ref. 212. The solid line represents the ideal KCl line.

1.5. Concluding remarks

In this study, where we benefit from the excellent thermal stability and wide liquid range of RTILs, the effect of different anions on the conductivity of highly pure [bmim]$^+$ RTILs has been investigated over the exceptionally wide temperature range of (248 to 468) K. Electrical conductivity measurements of the RTILs [bmim][DCA], [bmim][PF$_6$], [bmim][TfO] and [bmim][TA] have been performed, supplementing a companion study on the cation-dependence of κ for imidazolium RTILs. At a given temperature conductivity decreases in the order [bmim][DCA] > [bmim][TA] > [bmim][TfO] > [bmim][PF$_6$]. Temperature dependence is well described by the VFT equation. Whilst our data compare favorably with some literature results, significant deviations from others were noted. Walden plots of the molar conductance, available for [bmim][PF$_6$], [bmim][TfO] and [bmim][TA] in the limited temperature range of (278 to 338) K suggest that these RTILs can be classified as high-ionicity ionic liquids.

2. Microemulsions with an ionic liquid surfactant and room temperature ionic liquids as polar phase

2.1. Introduction

Ionic liquids (ILs), especially imidazolium and pyridinium based ionic liquids, have attracted more and more attention in recent years, because of their outstanding properties, as described in detail in section II.[32,33] Beside these ionic liquids, there is also a growing interest in protic ionic liquids[21,28] and research towards greener ILs.[195,223]

Particular attention has been paid to room temperature ionic liquids (RTIL). The RTIL ethylammonium nitrate has been described almost a century ago by Walden.[1] This polar, colorless liquid exhibits a melting point of 14°C[22] and is supposed to form three-dimensional hydrogen bonded networks.[24,224] Recent studies confirm this assumption with far-IR spectroscopy and static quantum chemical calculations.[225,226] The formation of amphiphilic association structures in and with ionic liquids, such as micelles, vesicles, microemulsions and liquid crystalline phases has been reviewed recently.[19,20,227] The formation of micelles formed by common surfactants in EAN was first documented in 1982 by Evans *et al.*,[22,24] whereas the formation of liquid crystals of lipids in EAN[224] as well as lyotropic liquid crystals of non-ionic surfactants were observed in the same RTIL.[228] The formation of liquid crystalline phases of binary mixtures of [C_{16}mim][Cl] and EAN has also been described lately.[31] The self-aggregation of common ionic and non-ionic surfactants in imidazolium based RTIL was also reported.[25,26,27]

As imidazolium based ionic liquids (ILs) with long-chain hydrocarbon residues exhibit surfactant properties in water,[29,229-233] Thomaier *et al.* could recently demonstrate that the surfactant like ionic liquid (SLIL) 1-hexadecyl-3-methylimidazolium chloride ([C_{16}mim][Cl]) forms colloidal structures in EAN as well. They found a critical aggregation concentration (cac) approximately ten times higher than the critical micelle concentration (cmc) in the corresponding aqueous system.[29] This gap is in accordance with the results of Evans *et al.*, who found that the cacs of classical surfactants in EAN are between 5 and 10 times higher than in water.[22] Starting from these anterior results that show clearly the possibility to use [C_{16}mim][Cl] in order to obtain colloidal structures in EAN, we decided to go a step beyond and formulate non- aqueous microemulsions containing [C_{16}mim][Cl] as surfactant and a RTIL as the polar phase. Microemulsions are thermodynamically stable, isotropic transparent

IV.2.1. Introduction

mixtures of at least a hydrophilic, a hydrophobic and an amphiphilic component. In common microemulsions the polar liquid is water or a brine solution.

Efforts have been made to replace water by a mixture of molten salts (nitrate mixtures of ethylenediamine/ammonia/potassium) in a system composed of sodium dodecyl sulfate (SDS), 1- pentanol and decane.[234] In recent years attempts have also been made to formulate nonaqueous reverse microemulsions containing RTILs as polar micro-environment. [160,192,235-238] Recently, microemulsions, where ILs are used as oil substitutes, water substitutes, additives or surfactants have been reviewed.[19,20] These investigations deal with non-polar solvents such as cyclohexane,[239] p-Xylene[237] and toluene[238] as continuous phase, a room temperature ionic liquid and a nonionic surfactant. Atkin et al. investigated recently microemulsions composed of nonionic alkyl oligoethylenoxide surfactants (C_iE_j), alkanes (octane, decane, dodecane and tetradecane) and EAN as polar phase.[192] They observed a single broad small- angle X- ray scattering (SAXS) peak like in aqueous systems, the Teubner- Strey model was used to analyze the scattering curves. Gao et al. used 1-butyl-3-methylimidazolium tetrafluoroborate ([bmim]][BF$_4$]) as polar compound in nonaqueous microemulsions. Freeze-fracture electron microscopy (FFEM) indicated a droplet structure similar to classical water-in-oil (w/o) microemulsions.[235] Using small angle neutron scattering (SANS) Eastoe and co-workers demonstrated for the same system a regular increase in droplet volume with added [bmim]][BF$_4$].[160] Li et al. investigated the micropolarity of microemulsions consisting of [bmim][BF$_4$], TX-100 and toluene.[238] By the addition of [bmim][BF$_4$] into the TX-100/toluene reverse micelles, the polarity of the IL/o microemulsions increased, after the formation of the IL pools the polarity remained constant.[89]

All studies concerning ILs in microemulsions described in literature have been performed below the boiling point of water. In the present study, we are interested in microemulsions that are stable over a wide temperature range at ambient pressure. For this purpose, water must be replaced by a RTIL. Most investigations dealing with microemulsions comprising RTIL as polar pool described in literature use nonionic surfactants. The main disadvantage of these systems is the reduced thermal stability and the high temperature sensitivity induced by the nonionic surfactants. To the best of our knowledge, other studies concerning microemulsions composed of oil, a cationic long chain ionic liquid surfactant (surfactant IL), cosurfactant and room temperature ionic liquid as polar microenvironment have not been reported in the open literature. We are interested in reverse microemulsions with nano- sized

RTIL polar cores exhibiting a thermal stability above the stability range of common microemulsions (>100°C). The microemulsions with surfactant IL exhibit different properties compared to the microemulsions with nonionic surfactants described in literature. All ingredients exhibit high boiling points and thermal stability, which predestines the microemulsions for high temperature applications.

2.2. Investigations at ambient temperature

2.2.1. Abstract

Nonaqueous microemulsions comprising an ionic liquid as surfactant and a room-temperature ionic liquid as polar pseudo-phase are presented.

Microemulsions containing the long- chain ionic liquid 1-hexadecyl-3-methyl-imidazolium chloride ([C_{16}mim][Cl]) as surfactant, decanol as cosurfactant, dodecane as continuous phase and room temperature ionic liquids (ethylammonium nitrate (EAN) and 1-butyl-3-methylimidazolium tetrafluoroborate ([bmim]][BF_4]), respectively) as polar micro-environment have been formulated.

Droplet structures of EAN/o have been confirmed with conductivity measurements. In presence of EAN a model of dynamic percolation could be applied. Dynamic light scattering (DLS) measurements indicated a swelling of the formed nano-structures with increasing amount of EAN, a linear dependence of the hydrodynamic radii on the EAN weight fraction was observed. Both systems exhibit a single broad peak in SAXS and follow a characteristic q^{-4} dependence of the scattering intensity at large q values. The Teubner-Strey model was successfully used to fit the spectra giving f_a, the amphiphilic factor, and the two characteristic lengths of microemulsions, namely the periodicity, d, and the correlation length, ξ. Furthermore, the specific area of the interface could be determined from the Porod limit and the experimental invariant. The amphiphilic factor clearly demonstrated structural differences between the two systems confirming that the nature of the polar ionic liquid plays an important role on the rigidity of the interfacial film. The adaptability of three different models ranging from spherical ionic liquid in oil over repulsive spheres to bicontinuous structures has been checked.

IV.2.2. Investigations at ambient temperature

2.2.2. Sample handling and experimental path

The synthesis of the ionic liquids used here are described in section III.2. ILs were stored in a N_2 filled glove box to avoid any contact with water and air.

The pseudo ternary phase diagrams were recorded at 30.0 ± 0.5°C, because of the larger single phase region observed compared to 25°C, especially in the case of [bmim][BF$_4$]. The phase determination was done according to the dynamic and static procedure described by Clausse et al.[240] The realms-of-existence of a clear and monophasic solution of the quaternary system were determined as a pseudo- ternary mixture implicating a pseudo constituent formed by the clear and homogenous solution of [C$_{16}$mim][Cl]/decanol in a molar ratio of 1:4. The molar ratio of [C$_{16}$mim][Cl]/decanol was chosen 1:4 because with this ratio the largest single-phase area was observed. In tubes different proportions of the surfactant-cosurfactant mixture to dodecane were prepared in 5% steps between 0 and 100 weight percent of the surfactant-cosurfactant solution. The RTIL was added under a nitrogen flow drop by drop under gentle agitation. The phase transitions were observed through cross polarized filters. The weight fractions of RTIL at which transparency-to-turbidity occurred were derived from precise weight measurements. The composition of the solution was then calculated from the masses of surfactant-cosurfactant, dodecane and RTIL. It was estimated that the accuracy of the measurements for the transparency-to-turbidity transition was better than ± 2%.

All measurements were carried out along an experimental path, where the amount of surfactant and cosurfactant was kept constant and the amount of RTIL was increased. The experimental paths were chosen in relation with the topology of the partial phase diagrams with the intention that a high amount of RTIL can be added with a relatively low amount of surfactant used. The paths that are not identical for the two RTILs due to the different phase diagram toplogy, are described in results and discussion section.

2.2.3. Results and discussion

The most important equations used for the data evaluation are for a better overview repeated here, although they have already been reported in chapter II.

2.2.3.1. Microregions and phase behavior

The pseudo-ternary phase diagrams of the dodecane/([C$_{16}$mim][Cl]/decanol)/EAN and the corresponding dodecane/([C$_{16}$mim][Cl]+decanol)/[bmim][BF$_4$] systems at 30°C are illustrated in Figure IV-6.

Dodecane/([C₁₆mim][Cl]+decanol)/EAN phase diagram (Figure IV-6 a):

A large monophasic region (L) extending from oil rich- to EAN rich- regions could be observed. The boundaries between the microemulsion phase L and the multi-phase region are irregular and exhibit an appendage towards 100% EAN. Such irregular boundaries with an appendage were already reported for the system 1-hexanol/water/Brij30, where the appendage turns to 100% of 1-hexanol.[241] This observation gives first hints on the reduced ionic character of [C₁₆mim][Cl]. The partial nonionic character of the surfactant can be related to the smeared charge of the surfactant headgroup. The unmixing limit of EAN and the pseudo-constituent is 72% (w/w) of EAN. As observed with nonionic surfactants, such phase diagram topologies correspond to microemulsion systems showing a flexible interfacial film.

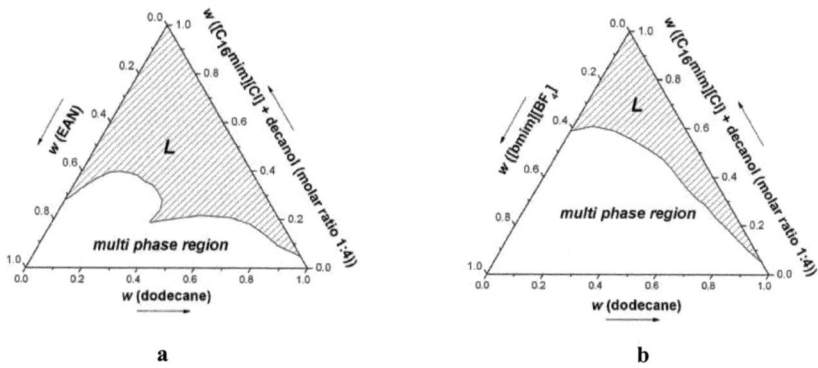

a b

Figure IV-6. Pseudo ternary phase diagram of (a) dodecane/([C₁₆mim][Cl]+decanol)/ EAN and (b) dodecane/([C₁₆mim][Cl]+decanol)/[bmim][BF₄] (in weight fraction at 30°C with a [C₁₆mim][Cl]/decanol molar ratio of 1 to 4). L represents the clear and isotropic single phase region.

Dodecane/([C₁₆mim][Cl]+decanol)/[bmim][BF₄] phase diagram (Figure IV-6 b):

The clear and isotropic 1-phase region (L) in the [bmim][BF₄] system is significantly smaller compared to the EAN system. The amount of [bmim][BF₄] that can be solubilized increases linearly with the surfactant/cosurfactant concentration until around 60% (w/w). Increasing [bmim][BF₄] concentration above its solubility limit leads the system to expel [bmim][BF₄] as

IV.2.2. Investigations at ambient temperature

a second liquid phase. This behavior is typical of rigid film microemulsions.[242] The phase diagram is regular, no appendage could be observed. The unmixing limit of [bmim][BF$_4$] and the pseudo- constituent is 41 wt% of [bmim][BF$_4$] which is one reason for the reduced single-phase area (L) compared to the EAN phase diagram.

2.2.3.2. Conductivity

Conductivity measurements can yield information about the percolative[240] or anti-percolative behavior of the microemulsions. For percolative systems the conductivity of w/o microemulsions shows a pronounced change over some orders of magnitude above a certain threshold.[128] At low water contents the conductivity is very low as oil is the continuous phase with reverse micelles well separated from each other. At the percolation threshold the droplets come sufficiently close to each other or form droplet clusters that an effective transport of charge carriers can take place.

For the microemulsion formed with EAN, an experimental path, where the amount of surfactant plus cosurfactant was kept constant at 30 wt% (P_S = 30), while the amount of EAN increases. Along this experimental path, a percolation phenomenon as described for aqueous microemulsions was observed. The specific conductivity as a function of the EAN volume fraction ϕ_{EAN} indicating the percolation phenomenon is illustrated in Figure IV-7.

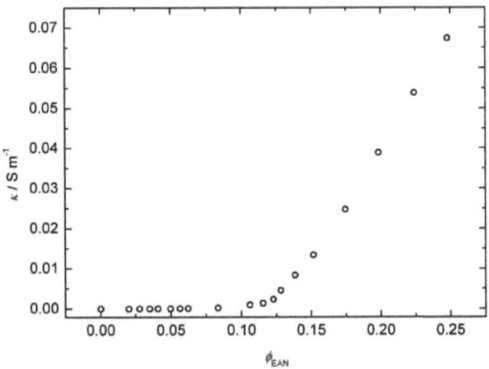

Figure IV-7. Specific conductivity, κ, versus ϕ_{EAN} for an experimental path with P_S = 30.

One of the conditions to observe a percolative behavior in a microemulsion system is that the interfacial film has to be flexible enough. For rigid film systems anti-percolation behavior is more likely to happen. The so-called percolation threshold ϕ_P^{EAN} where a significant change in conductivity was observed was determined following the method of Lagourette et al.[243] by plotting the conductivity $\kappa^{\%}$ versus the EAN volume fraction ϕ_{EAN}. The middle part of the curve can be fitted by a straight line whose intersection with the ϕ_{EAN} axes provides ϕ_P^{EAN} which is found to be 0.105. The droplet volume fraction can be written as

$$\phi = \frac{V_{EAN} + V_{[C_{16}mim][Cl]+decanol}}{V_{EAN} + V_{[C_{16}mim][Cl]+decanol} + V_{dodecane}} \qquad (IV-2)$$

with the assumption of ideal mixing, where V_i is the volume of compound i.

The percolation threshold ϕ_P in terms of the droplet volume fraction was determined to 0.33. At low EAN content ($\phi_{EAN} < 0.04$) the conductivity remains around (10^{-6}-10^{-5}) S m^{-1} and is 3-4 orders of magnitude lower than above the percolation transition. However, the conductivity is still much higher compared to a typical non-polar solvent. Eicke et al. described this fact for classical w/o microemulsions and explained this phenomenon by the assumption that nanodroplets carry positive or negative excess charges.[244] At higher EAN ($\phi_{EAN} > 0.11$) content, the conductivity increases linearly with the volume fraction of EAN. The model of dynamic percolation[138,139] assumes spherical independently moving droplets as in the charge fluctuation model. The conductivity can be described below and above the percolation threshold by appropriate asymptotic power laws,[245] where κ_1 and κ_2 correspond to the conductivities of the percolating droplets and of the homogeneous phase.

$$ln(\kappa) = ln(\kappa_1) + \mu \, ln\left(\left|\phi - \phi_P\right|\right) \qquad \phi >> \phi_p + \Delta \qquad (IV-3)$$

$$ln(\kappa) = ln(\kappa_2) - s \, ln\left(\left|\phi - \phi_P\right|\right) \qquad \phi << \phi_p - \Delta \qquad (IV-4)$$

Δ is the width of the transition region and μ and s are two characteristic exponents and can be determined from the slopes of equations IV-3 and IV-4. According to computer simulations the exponents μ and s vary from $\mu = 1.94$ and $s = 1.2$ for dynamic and $\mu \approx 2$ and $s \approx 0.6$-0.7 for static percolation. Consequently, it is possible to differentiate between dynamic and static percolation of microemulsions.

IV.2.2. Investigations at ambient temperature

Therefore, $\ln(\kappa)$ was plotted as a function of $\ln(|\phi - \phi_P|)$ in order to determine the critical exponents μ and s (Figure IV-8). The critical exponent μ could be obtained by fitting a straight line to the experimental data in the upper part of the diagram. The average value of parameter μ was 1.65, which differs from the expected theoretical value. Such differences between the experimental and the theoretical value have already been observed in literature.[245,246] As the exponent μ is almost the same for these two models, it cannot be used for a clear differentiation between the theoretical values. As can be seen in Figure IV-8, the lower curve is not linear over the whole range, and such behavior has also been described in literature before.[246] In that case, s was determined to be 1.28, which is in better agreement with the model of dynamic percolation. However, this is only a rough estimation, because the slope of the curve determining s depends on the number of points that are taken into account.

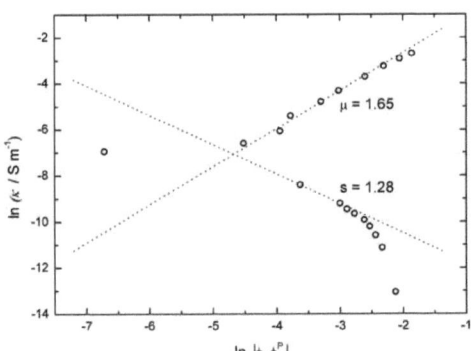

Figure IV-8. Determination of the critical exponents for the microemulsions with EAN.

According to the charge fluctuation model of Eicke et al.[128] the droplet radii r_d can be determined according to equation IV-5.

$$\kappa = \frac{\varepsilon\, \varepsilon_0\, k_B\, T\, \phi}{2\, \pi\, \eta\, r_d^3} \qquad (IV\text{-}5)$$

where ε_0 is the electric field constant, ε_r the relative permittivity of the continuous phase, ϕ the droplet volume fraction, T the temperature, η the viscosity of the continuous phase and k_B

the Boltzmann constant. To ensure that the charge fluctuation model is valid, ϕ values far below the percolation threshold in the range of 0.24 to 0.28 were used with T = 303 K, η = 0.00124 Pas[152] and ε_r = 2.2. The mean value of the droplet radii r_d was found to be 18.9 Å. It was described in literature that Eicke's model gives too large values of the droplet radii for droplets smaller than 70 Å.[130,131,132]

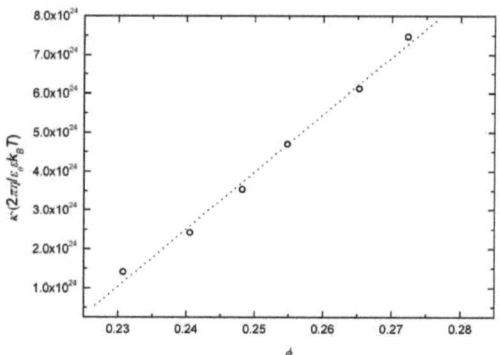

Figure IV-9. Determination of the droplet radii according to the charge fluctuation model.

In the case of the microemulsions formed with [bmim][BF$_4$] conductivity was measured along an experimental path where the amount of surfactant+cosurfactant was kept at 65 wt%. This relatively high amount of surfactant was necessary to obtain single phase microemulsions as can be seen from the phase diagram.

IV.2.2. Investigations at ambient temperature

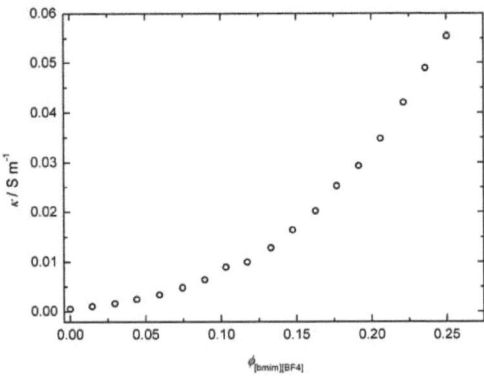

Figure IV-10. Specific conductivity, κ, versus $\phi_{[bmim][BF_4]}$ for an experimental path where the amount of surfactant+cosurfactant was kept at 65 wt%.

No significant change in slope of the curve could be found, the conductivity increases continuously with increasing amount of [bmim][BF$_4$] as can be seen from Figure IV-10. Consequently, no percolation threshold could be determined. This effect could be related to the high surfactant amount used resulting in a higher electric conductivity without the addition of RTIL. Furthermore, these results yield important information about the structure of the microemulsions. The conductivity where only very small amount of RTIL was added was already about two orders of magnitude higher compared to the EAN system. This can be interpreted that no well separated droplets of [bmim][BF$_4$] with dodecane as continuous phase are present. Nevertheless, the viscosity as will be discussed in the following section is relatively low, which speaks against the formation of a bicontinuous structure.

2.2.3.3. Viscosity

Along the experimental paths at 30°C dynamic viscosities, η, of the microemulsions show a Newtonian behavior. The mean values of η versus the volume fraction of RTIL are given in Figure IV-11 for the EAN system and the [bmim][BF$_4$] system, respectively. For both systems, no significant change in slope could be detected. The difference in viscosities of the continuous phase and the RTIL is not large enough to detect a significant effect.

IV. Results and discussion

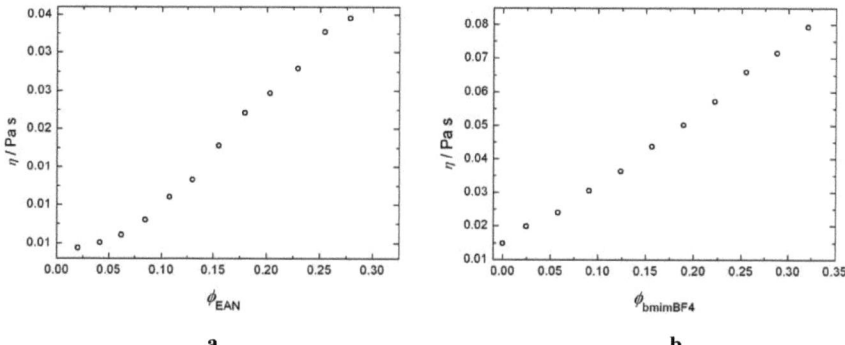

Figure IV-11. Dynamic viscosities at 30°C in dependence of the volume fraction of EAN a) and [bmim][BF$_4$] b), respectively.

Hence, it was not possible to confirm the dynamic percolation behavior of the EAN-system quantitatively by the application of the analogous power laws, because the viscosity did not increase over several orders of magnitude as in the conductivity measurements. The viscosity of the [bmim][BF$_4$] system changes almost linearly with the [bmim][BF$_4$] volume fraction.

2.2.3.4. Dynamic light scattering (DLS)

In order to get more information about the size and shape of the IL/o microemulsion droplets, dynamic light scattering measurements have been performed as a function of the amount of RTIL. The same experimental path as used for the previous experiments was chosen. The amount of surfactant and cosurfactant $w([C_{16}mim][Cl]+decanol)$ was kept constant at 30 weight percent for the EAN system and 65 weight percent for the [bmim][BF$_4$] system, respectively. Along the experimental path the amount of RTIL was increased. For the [bmim][BF$_4$] system the scattering intensity was too low for a convenient data evaluation, this is likely to be linked to the high surfactant/co-surfactant concentration used. For the EAN microemulsion the autocorrelation functions $(g^{(2)}(t)-1)$ versus the lag times τ showed a monomodal trend and could well be described by a single exponential fit. To evaluate the size of the droplets, the apparent hydrodynamic radii (R_{Happ}) were calculated according to the Stokes- Einstein equation. We assumed dodecane to be the continuous phase of the microemulsion and used $\eta = 0.00124$ Pas[152] and T = 303 K. The droplets show a swelling in

IV.2.2. Investigations at ambient temperature

dependence of the addition of the polar component EAN Figure IV-12 shows the hydrodynamic radii as a function of ϕ_{EAN} indicating a swelling of the formed structures.

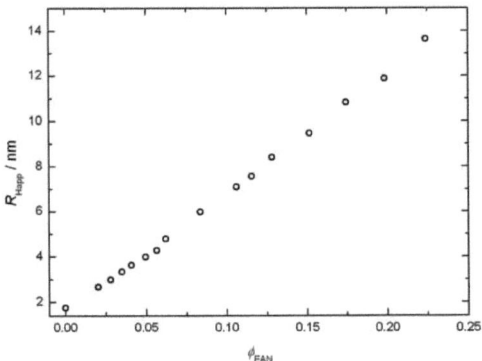

Figure IV-12. Hydrodynamic radii R_{Happ} in dependence of ϕ_{EAN} at 30°C indicating a swelling of the reverse micelles with increasing amount of EAN.

2.2.3.5. Small angle X-ray scattering (SAXS)

All SAXS experiments have been carried out along the same experimental path described for the previous experiments.

It has been shown in literature that small angle X-ray spectra in reverse microemulsions containing water, oil and surfactant exhibit a single broad peak.[159,161,162] Recently, microemulsions consisting of nonionic surfactant, oil and EAN were investigated with SAXS. These microemulsions showed also a single broad small-angle X-ray scattering peak similar to aqueous systems.[192]

The specific area Σ (cm^2/cm^3) at the polar/non-polar interface can be obtained from the SAXS spectra in the large q-range where the scattered intensity follows a q^{-4} behavior. This so-called Porod regime can be described by equation IV-6, where $\Delta\rho$ is the scattering length density difference between the polar and the non-polar part.[247]

$$\lim_{q \to \infty} (I\, q^4) = 2\pi \Delta\rho^2\, \Sigma \qquad (IV\text{-}6)$$

This regime is only obtained, if a thin interface separates two media of different scattering length densities.[247] Such a hypothesis is verified by evaluating the so-called experimental invariant (Q_{exp}, eq. IV-7), which is a general property of scattering spectra and is calculated from the scattering intensity.[170] Furthermore, an additional term was added in order to consider the contribution to the integral at higher q- values (eq. IV-7).[170]

$$Q_{exp} = \int_0^\infty I(q) q^2 \, dq = \int_0^{q_{exp}} I(q) q^2 \, dq + \frac{[I(q) q^4]}{q_{exp}} \qquad \text{(IV-7)}$$

Theoretically, Q_{exp} must be equal to the theoretical invariant (Q_{theo}, eq. IV-9) calculated with the volume fraction of the polar compounds Φ_{pol} (eq. IV-8) and the scattering length density contrast $\Delta\rho$.

$$\Phi_{pol} = \Phi_{RTIL} + \Phi_{OH} + \Phi_{[mim][Cl]} \qquad \text{(IV-8)}$$

$$Q_{theo} = 2\pi^2 \Delta\rho^2 \Phi_{pol} (1 - \Phi_{pol}) \qquad \text{(IV-9)}$$

For the present microemulsions the compositions of the ionic liquid and organic pseudo-phases within the microemulsion, which are required to calculate $\Delta\rho$, are not known. In particular the distribution of the cosurfactant in the film and the organic phase, repectively, is unknown. This is a recurrent problem for modeling microemulsion systems. Sometimes it is possible to solve this problem when microemulsions in equilibrium with an aqueous and an oil phase are studied.[248] In that case, a careful chemical analysis of the different phases in equilibrium with the microemulsion enables the determination, by using mass action laws, of the exact composition of the organic and aqueous phases within the microemulsion as well as the composition of the interface.

The specific surface, Σ, can be determined by combining equations IV-6 - IV-9 yielding equation IV-10. The only assumption which is made here, is that $Q_{exp} = Q_{theo}$. However, the evaluation of Q_{theo} is difficult because the exact distribution of the components is not known. As the background and the experimental invariant are precisely determined, Q_{exp} is the more reliable value.

IV.2.2. Investigations at ambient temperature

$$\Sigma = \frac{\lim_{q \to \infty}(I\, q^4)\, \pi\, \Phi_{pol}\,(1-\Phi_{pol})}{\int_0^{\infty} I(q)\, q^2\, dq} \tag{IV-10}$$

The Porod's limit and the experimental invariant were both determined experimentally from the spectra, and Φ_{pol}, the volume fraction of the scattering part, i.e. the polar reverse micelles, was evaluated from the weighted amounts of the different ingredients and their densities. As the studied ILs and the surfactant are sparingly soluble in the alcane phase we can consider with good confidence that along the experimental path the polar IL (EAN or [bmim][BF₄]) and the polar heads of the surfactant –[mim][Cl] are inside the reverse micelles. Concerning decanol, as it is cosoluble with dodecane and insoluble in the polar ILs, a partitoning of the hydroxyl group of the decanol occurs between the alkane phase and the alkane/polar IL interface. For that reason, a parameter α was defined that takes the distribution of decanol between the interfacial film and the oil phase into account For $\alpha = 1$, the OH groups of decanol are completely concentrated in the interfacial film. For $\alpha = 0$, all decanol is in the alkane phase and does not contribute to the interfacial film. Φ_{pol} and Σ were calculated for different α values in 0.25 steps varying from 0 to 1. These results are summarized in Table IV-4 for the EAN systems and Table IV-5 for the [bmim][BF₄] systems, respectively.

Table IV-4. Σ for the EAN microemulsions along the experimental path for different α values with the corresponding invariant and Porod limit.

ϕ_{EAN} / %			0.00	2.07	4.24	6.19	10.47	15.17
$\phi_{dodecane}$ / %			73.36	70.94	68.47	66.19	61.19	55.92
$\phi_{decanol}$ / %			17.78	18.02	18.22	18.44	18.92	19.30
$\phi_{[C16mim][Cl]}$ / %			8.86	8.97	9.07	9.18	9.42	9.61
Q_{exp} / (cm^{-4}/10^{21})			1.57	1.26	1.61	1.66	1.68	1.87
Porod limit / (cm^{-5}/10^{27})			22.0	10.0	10.0	8.9	8.3	7.3
Σ / (cm²/cm³/10⁶)		$\alpha = 0.00$	1.30	1.22	1.33	1.42	1.84	1.85
		$\alpha = 0.25$	1.50	1.32	1.41	1.49	1.90	1.89
		$\alpha = 0.50$	1.69	1.43	1.49	1.55	1.95	1.93
		$\alpha = 0.75$	1.88	1.54	1.57	1.62	2.01	1.96
		$\alpha = 1.00$	2.07	1.64	1.65	1.68	2.06	2.00

IV. Results and discussion

Table IV-5. Σ for the [bmim][BF$_4$] microemulsions along the experimental path for different α values with the corresponding invariant and Porod limit.

		0.00	2.28	4.42	6.41	11.12	16.10
$\phi_{\text{[bmim][BF4]}}$ / %							
ϕ_{dodecane} / %		38.77	35.84	32.95	30.03	23.70	16.86
ϕ_{decanol} / %		40.87	41.31	41.81	42.43	43.52	44.76
$\phi_{\text{[C16mim][Cl]}}$ / %		20.36	20.57	20.81	21.13	21.66	22.28
Q_{exp} / (cm^{-4}/10^{21})		4.57	4.80	5.03	5.40	5.66	5.82
Porod limit / (cm^{-5}/10^{28})		5.0	5.0	5.0	5.0	5.0	5.0
Σ / (cm^2/cm^3/10^6)	α= 0.00	2.24	2.78	3.20	3.43	4.19	4.89
	α= 0.25	2.56	3.06	3.47	3.67	4.39	5.05
	α= 0.50	2.87	3.35	3.72	3.90	4.59	5.21
	α= 0.75	3.17	3.62	3.97	4.12	4.77	5.36
	α=1.00	3.47	3.89	4.21	4.33	4.95	5.50

With increasing α, Σ increases for a constant volume fraction of RTIL. Moreover, Σ decreases with the amount of EAN and increases slightly with the amount of [bmim][BF$_4$]. Σ as a function of the RTIL volume fraction can be seen in Figure IV-13 a for EAN and Figure IV-13 b for the [bmim][BF$_4$] systems, respectively.

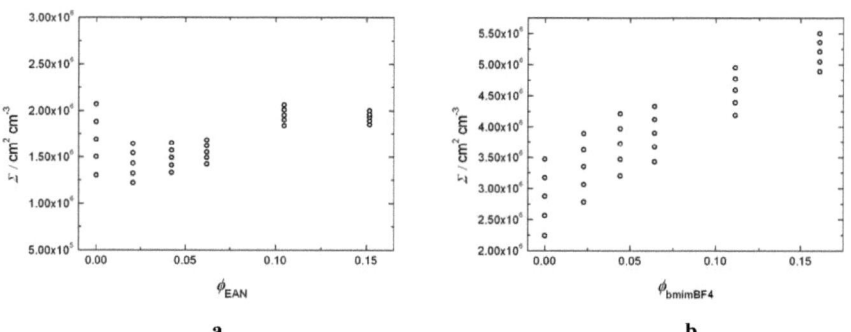

Figure IV-13. Σ as function of the RTIL volume fraction for α values between 0 and 1 (0.00; 0.25; 0.50; 0.75; 1.00), α increases from the bottom up.

IV.2.2. Investigations at ambient temperature

Furthermore, the SAXS spectra of both systems were fitted by the Teubner-Strey (TS) model,[161,249] which can be used to describe small angle scattering by microemulsions. A characteristic q^{-4} dependence of the intensity distribution at large q values can be observed. The TS model comprises three fitting parameters to describe the broad peak and the q^{-4} decay. The scattering function can be written as,

$$I(q) = \frac{1}{a_2 + c_1 q^2 + c_2 q^4} + I_{in} \qquad \text{(IV-11)}$$

with $a_2 > 0$, $c_1 < 0$, $c_2 > 0$ and the stability condition $4 a_2 c_2 - c_1^2 > 0$.

From the fit, two length scales characterizing microemulsions can be obtained: ξ and d, where d represents the domain size and ξ a correlation length. The two length scales can be expressed as:[161]

$$\xi = \left[\frac{1}{2} \left(\frac{a_2}{c_2} \right)^{1/2} + \frac{1}{4} \frac{c_1}{c_2} \right]^{-1/2} \qquad \text{(IV-12)}$$

$$d = 2\pi \left[\frac{1}{2} \left(\frac{a_2}{c_2} \right)^{1/2} - \frac{1}{4} \frac{c_1}{c_2} \right]^{-1/2} \qquad \text{(IV-13)}$$

The amphiphilic factor f_a[167-169] can be calculated according to:

$$f_a = \frac{c_1}{(4 a_2 c_2)^{1/2}} \qquad \text{(IV-14)}$$

f_a equals to 1 for the disorder line, where the solution loses its quasiperiodical order.[249] The liquid crystalline lamellar phase corresponds to $f_a = -1$.[169] These values delimit the region, where microemulsions may be found.[168] The factor ranges between -0.9 and -0.7 for well-structured bicontinuous microemulsions.[167,169]

IV. Results and discussion

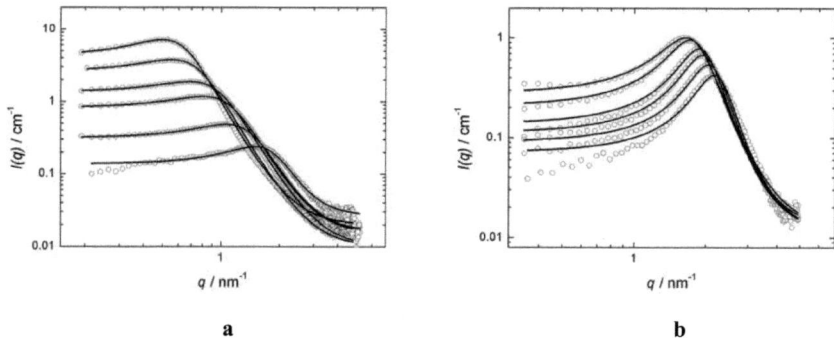

a   b

Figure IV-14. a) SAXS curves in dependence of the amount of EAN at 30°C (w_{EAN} = 0%, 3%, 6%, 9%, 15%, 21%), the scattering intensity increases with w_{EAN}, full lines fit with the TS model. b) SAXS curves in dependence of the amount of [bmim][BF$_4$] at 30°C (w[bmim][BF$_4$] = 0%, 3%, 6%, 9%, 15%, 21%), the scattering intensity increases with w[bmim][BF$_4$], full lines fit with the TS model.

For both systems, the position of the peak maxima varies with the amount of RTIL and is shifted to smaller q values for increasing RTIL content. The scattering curves could be well described by the Teubner-Strey formula according to eq. IV-11 (full lines Figure IV-14). The SAXS curves for different RTIL concentrations and the corresponding fits from the TS model (full lines) are illustrated in Figure IV-14 a for the EAN Figure IV-14 b for the [bmim][BF$_4$] microemulsions.

However, for the [bmim][BF$_4$] system the data points at low q- values are not quantitatively described by the TS formula. In Table IV-6 the obtained values for the periodicity d, the correlation length ξ, and the amphiphilic factor f_a are summarized. For the EAN system, d and ξ increase with the amount of EAN, whereas the amphiphilic factor ranges between -0.55 and -0.69. This effect can be associated with the swelling of the formed structures when the volume fraction of EAN is increased.

For the microemulsion containing [bmim][BF$_4$] the correlation length ξ does not change significantly and the periodicity increases slightly, which is related to the high surfactant mass ratio used. The amphiphilic factor ranges between -0.85 and -0.93 which is very close to the limit of well structured bicontinuous microemulsions.

IV.2.2. Investigations at ambient temperature

Table IV-6. Periodicity *d*, correlation length ξ and amphiphilic factor f_a derived from Teubner-Strey fits in dependence of the composition of the microemulsion.

Microemulsion system	RTIL / wt%	ξ / nm	d / nm	d/ξ	f_a
EAN-dodecane-[C$_{16}$mim][Cl]+decanol	0	1.4	3.8	2.7	0.69
	3	1.6	5.1	3.1	0.60
	6	1.9	6.5	3.4	0.55
	9	2.2	7.6	3.5	0.53
	15	2.9	9.6	3.3	0.56
	21	3.7	11.2	3.0	0.63
[bmim][BF$_4$]-dodecane-[C$_{16}$mim][Cl]+decanol	0	2.3	2.9	1.2	0.93
	3	2.4	3.0	1.3	0.92
	6	2.5	3.2	1.3	0.92
	9	2.5	3.3	1.3	0.91
	15	2.3	3.6	1.5	0.89
	21	2.1	3.7	1.8	0.85

These results can be compared to different model curves:[157]

The cubic random cell model (CRC) for bicontinuous structure (valid when $0.18<\Phi<0.82$):[172]

$$\Sigma\, d = 6\, \Phi_{pol}\, (1-\Phi_{pol}) \qquad \text{(IV-15)}$$

Spheres of water in oil (w/o) or (IL/o), respectively:[173]

$$\Sigma\, d = 4.84\, \Phi_{pol}^{1/3} \qquad \text{(IV-16)}$$

Repulsive spheres:[157]

$$\Sigma d = 4.32\, \Phi_{pol}^{2/3} \qquad \text{(IV-17)}$$

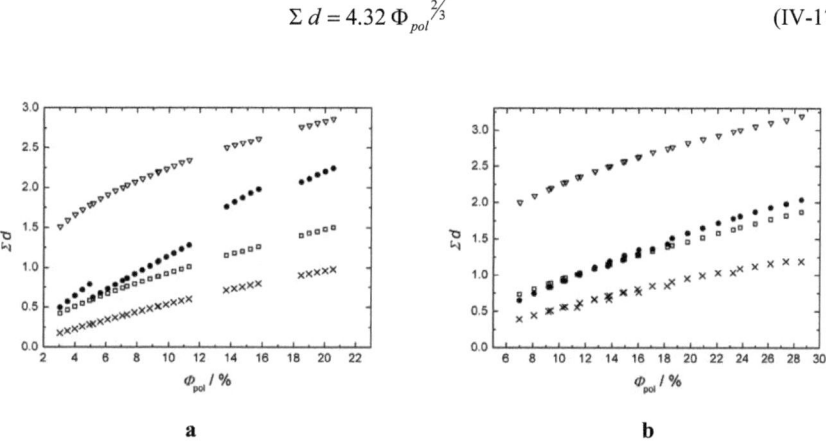

Figure IV-15. Σd versus Φ_{pol} for the EAN (a) and the [bmim][BF$_4$] (b) system of the experimental data ●, the cubic random cell model x, the model of IL/o spheres ∆ and the model of repulsive spheres □ for α between 0 and 1 (0.00; 0.25; 0.50; 0.75; 1.00).

For the EAN system, the experimental curve lies between the two sphere models confirming the existence of spherical structures, as shown in Figure IV-15 a. For the [bmim][BF$_4$] system, the experimental curve fits well with the model of repulsive spheres as shown in Figure IV-15 b.

2.2.4. Concluding remarks

Microemulsions consisting of dodecane, the ionic liquid surfactant [C$_{16}$mim][Cl], cosurfactant and RTIL (EAN and [bmim][BF$_4$], respectively, were formulated and investigated. The observed transparent and homogenous 1-phase region was characterized by different methods. All measurements were carried out along an experimental path, where the amount of surfactant and cosurfactant was kept constant and the amount of RTIL was increased. The experimental paths were chosen according to the topology of the phase diagrams and are hence not identical for the two systems. The conductivity along these experimental paths indicates a percolation behavior in the case of the EAN system, in the case of [bmim][BF$_4$] no significant change in slope could be observed. For the EAN microemulsion, a model of dynamic percolation could be applied. The hydrodynamic radii increase with the amount of

IV.2.2. Investigations at ambient temperature

EAN, confirming the swelling behavior of the formed structures. Both systems exhibit a single broad peak in SAXS, which is typical for microemulsions. Furthermore, a q^{-4} slope, characteristic of a sharp interface between the polar and the non-polar media could be observed. The specific area of the interface in dependence of RTIL volume fraction is in accordance with a swelling of the formed structures. The Teubner-Strey model was successfully used to fit the SAXS spectra. The SAXS data confirm the presence of spherical shape ionic liquid in oil microemulsions in the case of the EAN systems, whereas for [bmim][BF$_4$] systems the data suggest a probable contribution of [bmim][BF$_4$] to the formation of the interfacial film.

The striking difference in the interfacial rigidity of the microemulsions, which is induced by changing the nature of the polar IL, EAN and [bmim][BF$_4$], is confirmed by independent techniques: phase diagram determination, Σd vs. ϕ plots obtained from SAXS spectra treatments, conductivity and DLS measurements. Note that the experimental paths for the two systems are not identical and can also play an important role on the different behavior. The difference between the two systems may be linked to the change in the cohesive forces of the polar ILs or to the co-surfactant property of [bmim][BF$_4$], leading to a reinforcement of the interfacial film.

All ingredients show a good thermal stability and high boiling points, which is of particular interest for the formulation of high temperature stable microemulsions. The stability range of conventional microemulsions is limited to 100°C. An extension of the conventional thermal stability range of microemulsions opens a wide field of potential applications such as extractions or size controlled high temperature synthesis of inorganic materials. The thermal stability of these systems will be discussed in the following sections.

2.3. Ethylammonium nitrate in high temperature stable microemulsions

2.3.1. Abstract

In the previous section, microemulsions composed of the room-temperature ionic liquid ethylammonium nitrate (EAN) as polar phase, dodecane as continuous phase and 1-hexadecyl-3-methylimidazolium chloride ([C_{16}mim][Cl]), an IL that acts as surfactant, and decanol as cosurfactant at ambient temperature have been characterized. In this chapter, the high thermal stability of these microemulsions at ambient pressure is demonstrated. Along an experimental path, no phase change could be observed visually within a temperature ranging from 30°C to 150°C. The microemulsions are characterized with quasi-elastic light scattering measurements at ambient temperature and temperature dependent conductivity and temperature dependent small angle neutron scattering (SANS) experiments (30-150)°C.

Conductivity measurements confirm the high thermal stability and highlight a percolation phenomenon even at 150°C. DLS measurements at ambient temperature indicate a swelling of the formed structures with increasing amount of EAN up to a certain threshold. The SANS experiments were performed below this threshold. The data evaluation of such concentrated systems like microemulsions is possible with the "Generalised Indirect Fourier Transformation" method (GIFT). The small angle scattering data were evaluated via the GIFT method. For comparison the model of Teubner and Strey (TS) which was often used to describe scattering curves of microemulsions was also applied. The GIFT method yields good fits throughout the experimental path, while the TS model gives relatively poor fits. Conductivity, light scattering and SANS results are in agreement with the existence of a L_2-phase along the whole investigated temperature range. Moreover, these results clearly demonstrate the possibility to formulate high temperature stable microemulsions at ambient pressure with ionic liquids.

2.3.2. Sample handling and experimental path

The dried RTILs were stored in a N_2-filled glovebox. N_2-protection was also maintained during all subsequent steps of the microemulsion preparation.

In the previous section, microemulsions composed of EAN, [C_{16}mim][Cl], dodecane and 1-decanol were investigated. At ambient temperature, we chose an experimental path, where the amount of surfactant plus cosurfactant was kept constant at 30 weight percent ($P_S = 30$).[195]

IV.2.3. Ethylammonium nitrate in high temperature stable microemulsions

By visual observation we noticed that a higher amount of surfactant plus cosurfactant is necessary for a thermal stability up to 150°C at ambient pressure. With increasing amount of surfactant, the thermal stability increased. Hence, the total amount surfactant + cosurfactant was kept constant at 40 weight percent ($P_S = 40$). All experiments described here follow this experimental path.

2.3.3. Results and discussion

2.3.3.1. Density

For the calculation of the volume fraction of the dispersed phase, ϕ, densities, ρ, were measured between 25°C and 150°C. The temperature dependent densities of EAN, dodecane and the surfactant+cosurfactant mixtures are illustrated in Figure IV-16.

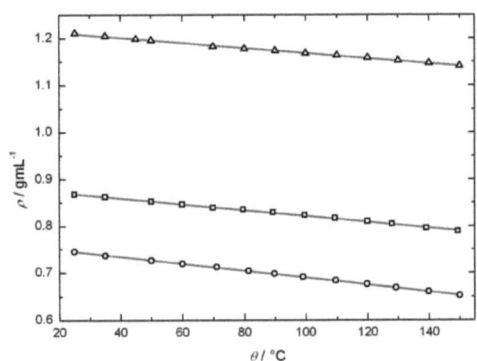

Figure IV-16. Temperature dependent densities of dodecane (○) [C$_{16}$mim][Cl]+decanol (molar ratio 1:4) (□) and EAN (Δ), full lines represent a linear fit.

The following linear density-temperature relationships were obtained:

ρ_{EAN} / g cm^{-3} = 1.223 - 0.00055·θ / °C

$\rho_{[C16mim][Cl]+decanol}$ / g cm^{-3} = 0.884 - 0.00063·θ / °C

$\rho_{dodecane}$ / g cm^{-3} = 0.763 - 0.00073·θ / °C

The temperature dependent volume fractions of the dispersed phase were calculated with the assumption of ideal mixing according to eq. III-8.

2.3.3.2. Visual observations

Visually, no phase change along the experimental path could be observed within 30°C and 150°C as illustrated exemplarily for 12 wt% EAN in Figure IV-17.

Figure IV-17. Photos of the microemulsion at 30°C and 150°C, indicating the excellent thermal stability of the systems (exemplary shown for w_{EAN} = 12 %).

2.3.3.3. Solubility of EAN in dodecane

Interfacial tension measurements of the EAN/dodecane interface indicate the immiscibility of the two compounds as the interfacial tension at 30°C between EAN and dodecane was determined to be 19.1 mN m^{-1}. The immiscibility of the polar phase EAN and the non-polar phase dodecane at ambient temperature and at 150°C was further confirmed by ^{1}H-NMR and refractive index measurements.

For this purpose, 5 g EAN were added to 5 g dodecane and stirred at ambient temperature and at 150°C, respectively, for three hours. Then, 2 mL of the upper phase (dodecane) were removed and cooled down to room temperature. By cooling down, no phase separation occurred. As the refractve indices of dodecane (n_D^{20} = 1.4221)[250] and EAN (n_D^{25} = 1.4524)[58] differ significantly, a partial miscibility of the compounds would be detectable. The refractive indices of pure dodecane and the oil phase of the dodecane-EAN mixture were identical within the uncertainty limits indicating the immiscibility of EAN in dodecane.

The phases were further characterized by means of ^{1}H-NMR measurements. Within the detection limits, EAN could not be detected with ^{1}H-NMR in the non-polar phase at both temperatures, 30°C and 150°C. The ^{1}H-NMR spectra of the pure EAN and pure dodecane are shown in Figure IV-18 and Figure IV-19, respectively.

IV.2.3. Ethylammonium nitrate in high temperature stable microemulsions

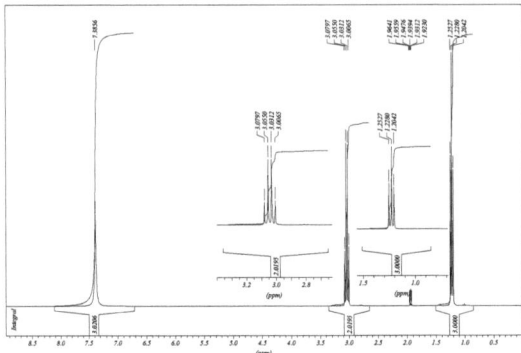

Figure IV-18. ^{1}H-NMR of EAN in acetonitrile-d$_3$.

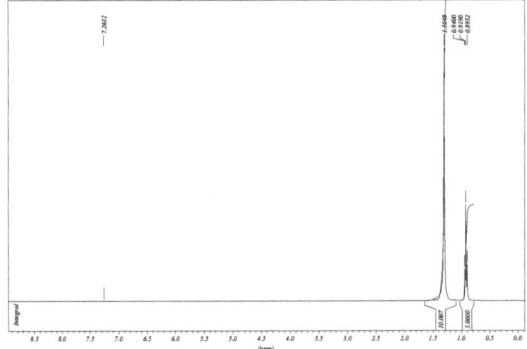

Figure IV-19. ^{1}H-NMR of dodecane in chloroform-d.

The upper phases (oil phases) are shown in Figure IV-20 for 30°C and Figure IV-21 for 150°C. No traces of EAN could be detected in the dodecane phase neither at 30°C, nor at 150°C. Therefore, we can assume with good confidence that EAN is not miscible in dodecane over the whole investigated temperature range.

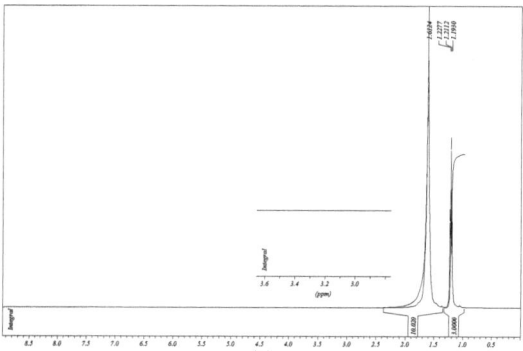

Figure IV-20. ^{1}H-NMR of dodecane saturated with EAN at 30°C, the quartet from CH_3-CH_2-NH_4^+ NO_3^- could not be detected demonstrating the immiscibility of EAN in dodecane.

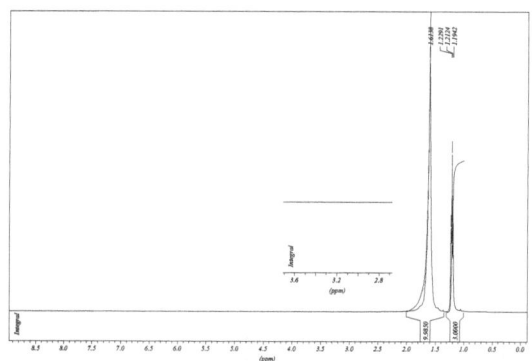

Figure IV-21. ^{1}H-NMR of dodecane saturated with EAN at 150°C, the quartet from CH_3-CH_2-NH_4^+ NO_3^- could not be detected demonstrating the immisciblity of EAN in dodecane.

2.3.3.4. Conductivity

Conductivity measurements have been performed at various temperatures between 30°C and 150°C (30°C, 60°C, 90°C, 120°C, and 150°C). The high temperature stable microemulsions show indeed a percolation phenomenon over the whole temperature range. The conductivity versus the droplet volume fraction at 30°C and 150°C are shown in Figure IV-22 a and b, respectively.

IV.2.3. Ethylammonium nitrate in high temperature stable microemulsions

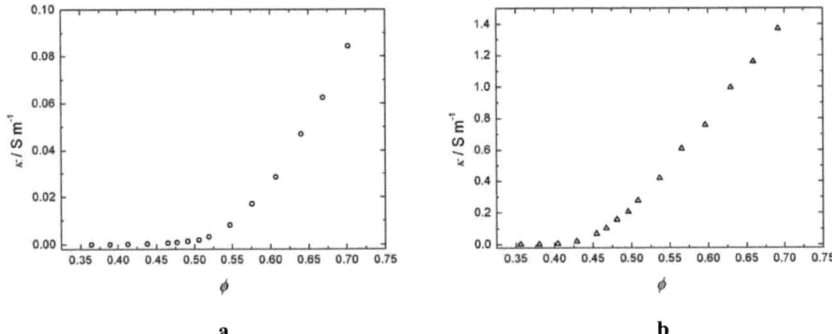

Figure IV-22. Specific conductivity as a function of the volume fraction ϕ at 30°C (a) and 150°C (b), respectively.

For both temperatures, the conductivity changes over several orders of magnitude along the experimental path demonstrating a percolation phenomenon. The specific conductivity at a fixed volume fraction increases with increasing temperature. For example the conductivity without the addition of EAN at $P_S = 40$ was determined to $6.69 \cdot 10^{-6}$ S m^{-1} at 30 °C and $9.80 \cdot 10^{-4}$ S m^{-1} at 150°C. At the maximum EAN content measured at $P_S = 40$ and $w_{EAN} = 36$ % the specific conductivity was determined to 0.08 S m^{-1} at 30°C and 1.37 S m^{-1} at 150°C.

Hence, two effects can be summarized, on the one hand the percolation phenomenon with increasing amount of EAN, on the other hand a significant increase in conductivity with increasing temperature at a fixed EAN content. Furthermore, the threshold, where the conductivity starts to increase sharply was shifted with increasing temperature to lower volume fractions. This observation was confirmed for different temperatures as can be seen in Figure IV- 23. From the insert of Figure IV- 23 one can see very clearly that the threshold, where the conductivity increases remarkably and percolation appears, is shifted with increasing temperature to the left hand side (lower ϕ).

This so called percolation threshold volume fraction ϕ_P was calculated from the inflection point of the curve $\log(\kappa) = f(\phi)$. The curves $\log(\kappa) = f(\phi)$ exhibit a sigmoid shape as shown in Figure IV-24. A fourth order polynomial was fitted to the curves, the percolation threshold volume fraction, ϕ_P, was calculated from the inflection point of the fit curve by setting the second derivative of the polynomial equal to zero.

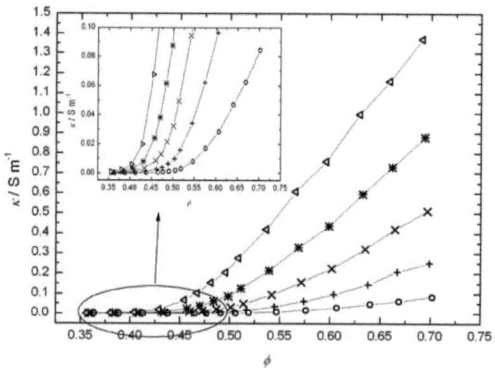

Figure IV- 23. Specific conductivity as a function of the volume fraction ϕ demonstrating the percolative behavior at different temperatures (30°C (○), 60°C (+), 90°C (×), 120°C (*) and 150°C(Δ)).

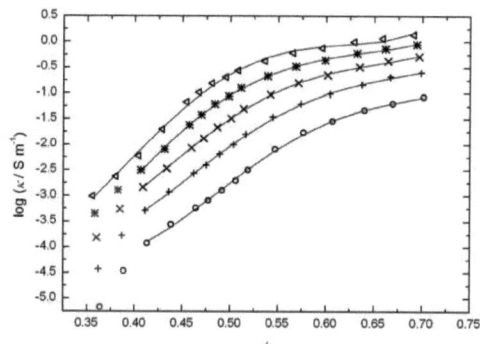

Figure IV-24. Determination of the percolation threshold volume fraction at various temperatures (30°C (○), 60°C (+), 90°C (×), 120°C (*) and 150°C(Δ)), full lines represent fourth order polynomial fits.

ϕ_p values as a function of temperature are summarized in Table IV-7 and are illustrated in Figure IV- 25. Interestingly, ϕ_p decreased linearly with increasing temperature. This effect can be explained by the increased motion of the formed IL nanodomains when the temperature is raised. Hence, the probability that two RTIL pools meet increases and the percolation threshold is thus shifted to smaller ϕ. The conductivity measurements demonstrate anyway the

IV.2.3. Ethylammonium nitrate in high temperature stable microemulsions

wide thermal stability range of the microemulsions.

Table IV-7. Percolation threshold volume fraction ϕ_P at investigated temperatures.

θ /°C	30	60	90	120	150
ϕ_P	0.48	0.46	0.43	0.41	0.39

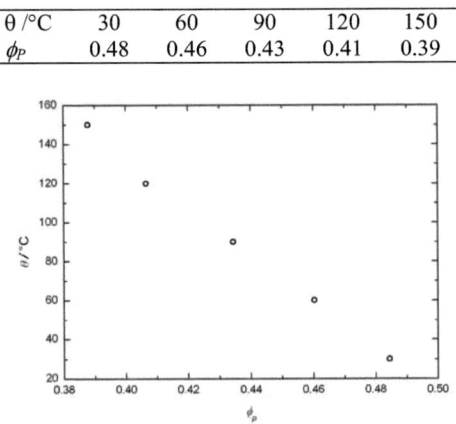

Figure IV- 25. Temperature dependence of the percolation threshold volume fraction ϕ_P.

Although the conductivity measurements yield hints of the microemulsion structure and the high thermal stability of these systems, they are no direct proof of the existence of microemulsions at high temperatures. Furthermore, information about the particle size is missing. Therefore, DLS measurements at ambient temperature have been performed.

2.3.3.5. Dynamic light scattering (DLS)

For the DLS measurements the amount of EAN was varied between 0 and 36 wt%. For EAN concentrations between 0% and 18% a single exponential decay of the autocorrelation function was observed. For higher EAN contents a bimodal decay was found. Exemplarily, the intensity correlation functions versus time are illustrated in Figure IV- 26 a and b for 12 and 30 wt% EAN, respectively. To each curve a single exponential decay (eq. IV-18) was fitted as it is shown by full lines in Figure IV- 26.

$$(g^{(2)} - 1) = a_0 + (a_1 \cdot \exp(-a_2 t))^2 \qquad \text{(IV-18)}$$

IV. Results and discussion

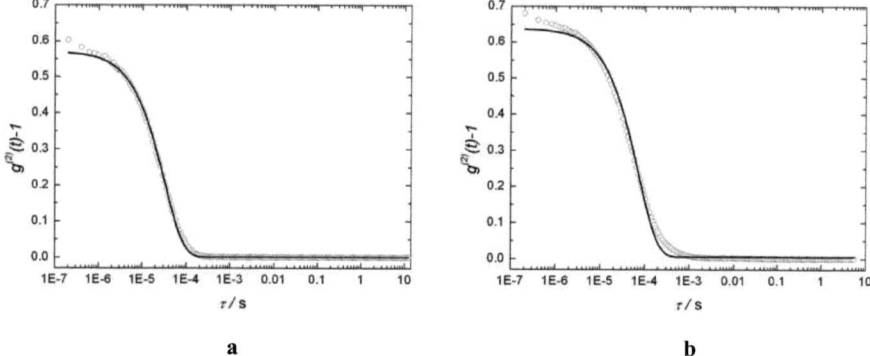

Figure IV- 26. Intensity autocorrelation functions at 30°C with $P_S = 40$ indicating a single exponential decay for $w_{EAN} = 12$ % (a) and a bimodal decay for $w_{EAN} = 30$ % (b), full lines are fits for a single exponential decay.

For 12 wt% EAN the intensity autocorrelation function could be well described by a single exponential decay. For 30 wt% EAN a single exponential decay was not sufficient to describe the curve. This gives first evidences for structural variations at higher EAN weight fractions. Interestingly, the percolation threshold is very close to the amount of EAN where a bimodal decay was observed. For the calculation of R_{Happ} we used the viscosity of dodecane (η = 1.236 cP)[152] at 30°C since it was assumed to be the continuous phase. By calculating the apparent hydrodynamic radii of the samples, where monomodal decays were observed (0 wt% EAN - 18 wt% EAN), a regular increase of the size with increasing amount of EAN was found. This regular swelling is shown in Figure IV-27. The polydispersity indices ranged between 0.24 and 0.30, no tendency in the variation of the polydispersity indices with increasing EAN content could be observed.

IV.2.3. Ethylammonium nitrate in high temperature stable microemulsions

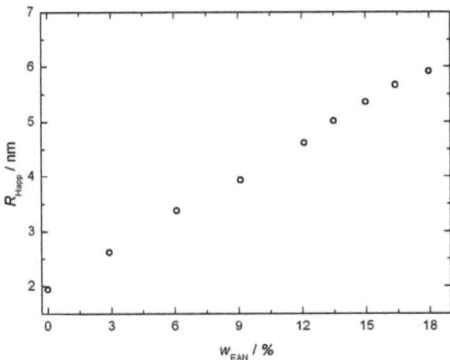

Figure IV-27. Regular swelling at 30°C with $P_S = 40$ of the reverse micelles with increasing EAN content for concentrations where a single exponential decay in the intensity autocorrelation function was observed.

To get more insight into the structure, particularly for those EAN weight fractions, where the intensity correlation function did not show a single exponential decay, a nonlinear data analysis provided by ALV was used. The data analysis program fits an integral type model function using a constrained regularization method. This regularization method (Contin) described by Provencher[148,149] is beside the non-negatively constrained least-squares method (NNLS)[150] the most often used one.[147] The following fit model was used, where G denotes the decay rate distribution function.

$$g^{(2)}(t) - 1 = \left(\int_{\Gamma_{min}}^{\Gamma_{max}} \exp(-\Gamma t) G(\Gamma) d\Gamma \right)^2 \qquad \text{(IV-19)}$$

In Figure IV-28 the distribution functions along the whole experimental path are illustrated. At low EAN content, the regular swelling is confirmed, above 18% EAN a second peak appears. One possible explanation for this second peak at higher amounts of EAN is the formation of droplet clusters or the formation of more elongated structures. Such hypotheses are in agreement with the conductivity measurements as above the percolation threshold elongated structures and channels are formed, at least droplets come sufficiently near to each other that an efficient transport of charge carriers is possible.

IV. Results and discussion

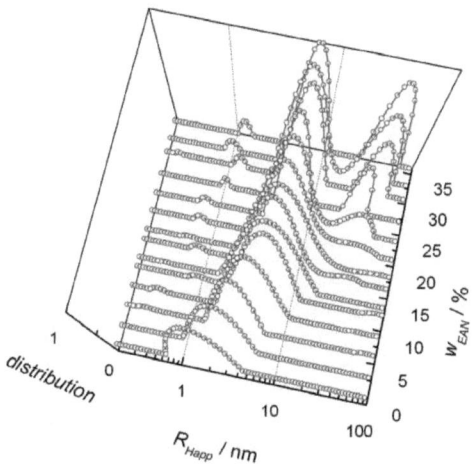

Figure IV-28. Size distribution in dependence of EAN content at 30°C with $P_S = 40$ indicating a swelling of the reverse micelles followed by the formation of percolated aggregates at $w_{EAN} > 18$ %.

The structural information obtained from the DLS measurements are not sufficient to prove any high thermal stability of the microemulsions. Hence, temperature dependent SANS experiments have been performed.

2.3.3.6. Small angle neutron scattering (SANS)

Small angle scattering (SAS) experiments are in general a powerful method to determine shape, size and internal structure of colloidal particles. They have been widely used to study aqueous microemulsions.[92,157,158,159] Small angle scattering spectra of aqueous microemulsions often exhibit a single broad correlation peak followed by a q^{-4} dependence at large q. As discussed in detail in the previous section, one model to describe small angle scattering curves from microemulsions is the TS formula, which was particularly developed for bicontinuous structures.

The temperature dependent SANS curves for 6 wt% EAN between 30°C and 150°C are shown in Figure IV-29. A single broad scattering peak was observed as often described for aqueous systems. At large q values the intensity exhibits as q^{-4} decay, but not over the whole range for large q values. Furthermore, the peak maxima became less pronounced with

IV.2.3. Ethylammonium nitrate in high temperature stable microemulsions

increasing temperature. Full lines in Figure IV-29 represent the fit obtained from the TS formula (eq. III-17).

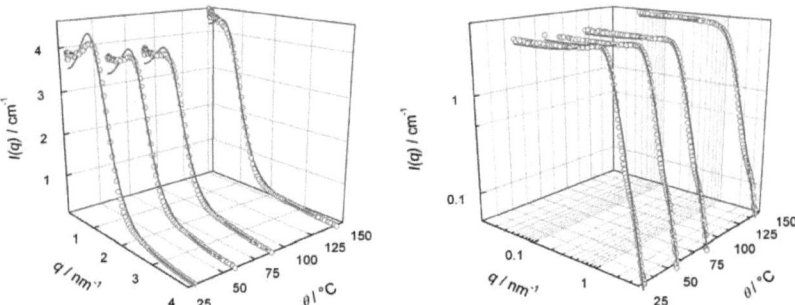

Figure IV-29. SANS spectra for 6 wt% EAN, P_S = 40 and 54 wt% dodecane d-$_{26}$ at different temperatures (30°C, 60°C, 90°C and 150°C), full lines are fit curves obtained from the TS formula.

Regarding the double linear plot it should be stressed that the peak maxima could not be described well with the TS formula. The double logarithmic plot indicates that the intensity does not follow a q^{-4} behavior over the whole large q range. As shown here exemplarily for 6 wt% EAN, the TS formula did not yield satisfactory fits. The same was found for a concentration series, particularly at 150°C, where the scattering peak is too weak to yield convenient fit results. These findings indicate that the present microemulsions do not show a bicontinuous structure. Such speculations are in agreement with the DLS and conductivity measurements that indicate the existence of reverse micelles with RTIL cores in a continuous oil matrix below the percolation threshold. Nevertheless, the domain size d, correlation length ξ and amphiphilic factor f_a give at least a first idea of the microemulsion structure and dimension. The TS fit was not applied for the SANS spectra at 150°C as the scattering peak was too weak to obtain a convenient fit. The parameters are summarized for a concentration series at different temperatures in Table IV-8.

Table IV-8. Fit results from the TS model with domain size, d, the correlation length, ξ, and the amphiphilic factor, f_a.

w_{EAN} / %	ξ /nm			d / nm			f_a		
	30°C	60°C	90°C	30°C	60°C	90°C	30°C	60°C	90°C
0.0	0.9	0.8	0.8	4.4	4.5	4.7	-0.24	-0.16	-0.12
2.9	1.1	1.0	1.1	4.9	5.0	5.2	-0.31	-0.25	-0.29
5.7	1.3	1.2	1.2	5.5	5.6	5.9	-0.36	-0.29	-0.25
9.4	1.6	1.5	1.5	6.2	6.2	6.6	-0.43	-0.38	-0.33
11.8	1.8	1.7	1.7	6.6	6.6	7.0	-0.48	-0.42	-0.37
15.7	2.0	18.9	18.7	6.9	7.0	7.3	-0.54	-0.49	-0.44

As the surfactant concentration along the experimental path is high, the particle interaction must be taken into account. For interacting globular particles the scattering intensity can be described as a product of the particle form factor $P(q)$ and the structure factor $S(q)$ that takes particle interaction into account. Another possibility to evaluate small angle scattering spectra of such concentrated systems is possible with the "generalized indirect Fourier transformation" (GIFT) method.[179,180,251,252] The GIFT method allows one to determine the form factor $P(q)$ and the structure factor $S(q)$ simultaneously, whereby the $P(q)$ is extracted in a model free way. Recent applications have shown that the GIFT method can also be used for polydisperse and nonglobular systems.[181,252] In the case of polydisperse interacting particles, the scattering intensity can be expressed according to eq. IV-21, where S_{eff} is the effective structure factor, P_{MF} the model free form factor and n the number density.

$$I(q) = n\, P_{MF}(q)\, S_{eff}(q) \qquad \text{(IV-20)}$$

The Percus-Yevick closure relation of hard sphere interaction yields a correlation peak similar to the TS model. It should be stressed that the Percus-Yevick effective structure factor is valid for uncharged systems. As we assume reverse micelles with EAN cores and dodecane as continuous phase, this assumption is reliable and was also confirmed with conductivity

IV.2.3. Ethylammonium nitrate in high temperature stable microemulsions

measurements. Further, the sizes of aggregates were assumed to be Schulz distributed, smearing effects from the wavelength distribution ($\Delta\lambda/\lambda = 10\%$)[207] were taken into account. S_{eff} is described by the parameters volume fraction ϕ, effective interaction radius R_{eff} and polydispersity. There is a kind of coupling between two of these parameters, namely polydispersity and ϕ. A too high polydispersity would lead to a too high volume fraction and vice versa. The calculation of an exact volume fraction is difficult in such complicated four component systems. Further, the solubility behavior of the ingredients changes as a function of temperature. Therefore, it is difficult to evaluate a convenient volume fraction. It is known that the polydispersity in microemulsions is high[151,253] and that the mean value is about 25%.[177] We fixed the polydispersity to 25% and assumed a Schulz distribution as it was done for aqueous microemulsions.[177] The polydispersity in bicontinuous systems can be higher, therefore concentrations below the percolation threshold at room temperature were chosen to ensure the existence of EAN droplets stabilized by surfactant with oil as continuous phase. The evaluation with the GIFT method for temperature dependent SANS curves for 6 wt% EAN between 30°C and 150°C yield excellent fit qualities as can be seen from Figure IV-30, full lines represent the fit curves.

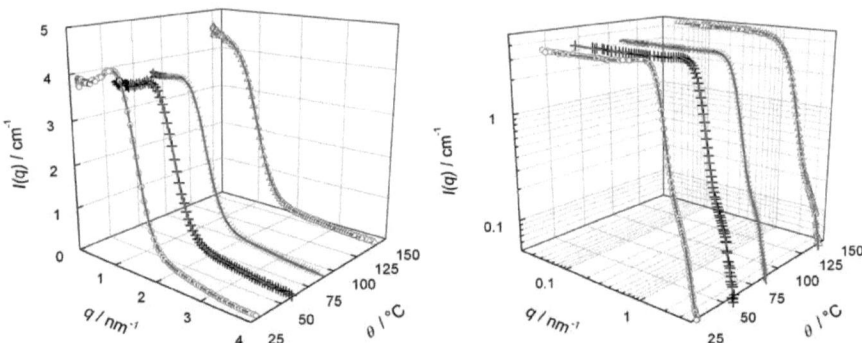

Figure IV-30. SANS for 6 wt% EAN, Ps = 40 and 54 wt% dodecane d-$_{26}$ at different temperatures (30°C (o), 60°C (+), 90°C (×) and 150°C (Δ)), full lines are fit curves from the GIFT evaluation using the Percus-Yevick effective structure factor.

The maxima r_{Ai}^{max} of the pair distance distribution functions $p(r)$ can be interpreted in terms of particle dimensions.[18] These functions are shown in Figure IV- 31. Interestingly, r_{Ai}^{max}

values do not change significantly between 30°C and 90°C (2.34 nm) and increase slightly for 150°C (2.54 nm). The $p(r)$ functions for the latter is broader compared to the $p(r)$ functions at lower temperatures. Nevertheless, there is no significant change in the shape of these functions with increasing temperature.

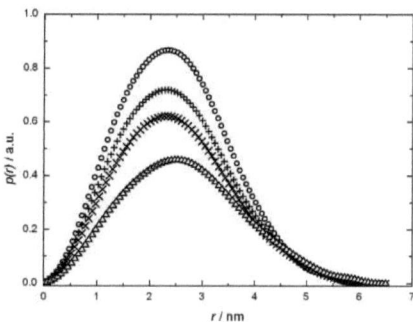

Figure IV- 31. Pair distance distribution functions $p(r)$, extracted from the GIFT method for 6 $wt\%$ EAN, P_S = 40 and 54 $wt\%$ dodecane d-$_{26}$ at different temperatures (30°C (○), 60°C (+), 90°C (×) and 150°C (Δ)).

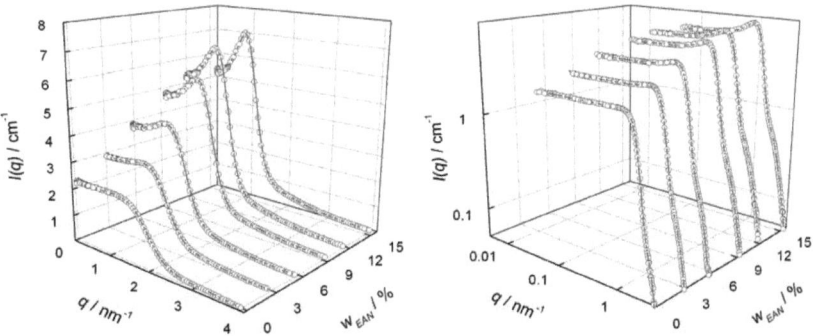

Figure IV-32. SANS curves along the experimental path for different EAN weight fractions (0 %, 3 %, 6 %, 9 %, 12 %, 16 %) at 30°C with P_S = 40, full lines are fit curves from the GIFT evaluation.

IV.2.3. Ethylammonium nitrate in high temperature stable microemulsions

In Figure IV-32, Figure IV- 33, Figure IV- 34 and Figure IV-35a concentration series along the described experimental path at 30°C, 60°C, 90°C and 150°C, respectively.

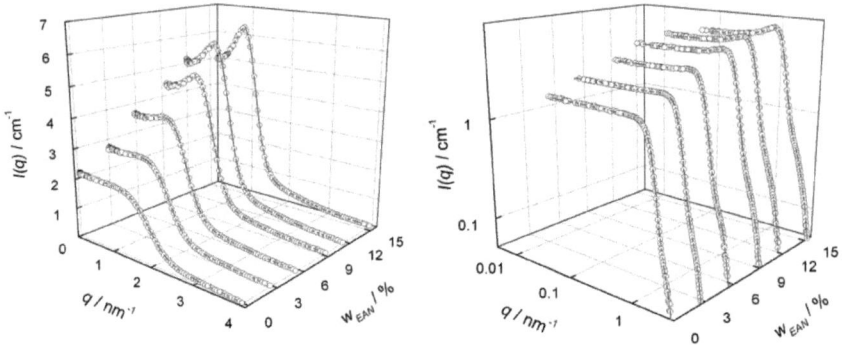

Figure IV- 33. SANS curves along the experimental path for different EAN weight fractions (0 %, 3 %, 6 %, 9 %, 12 %, 16 %) at 60°C with $P_S = 40$, full lines are fit curves from the GIFT evaluation.

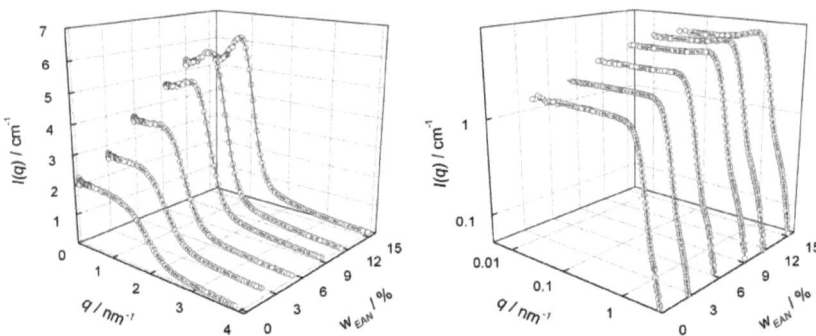

Figure IV- 34. SANS curves along the experimental path for different EAN weight fractions (0 %, 3 %, 6 %, 9 %, 12 %, 16 %) at 90°C with $P_S = 40$, full lines are fit curves from the GIFT evaluation.

IV. Results and discussion

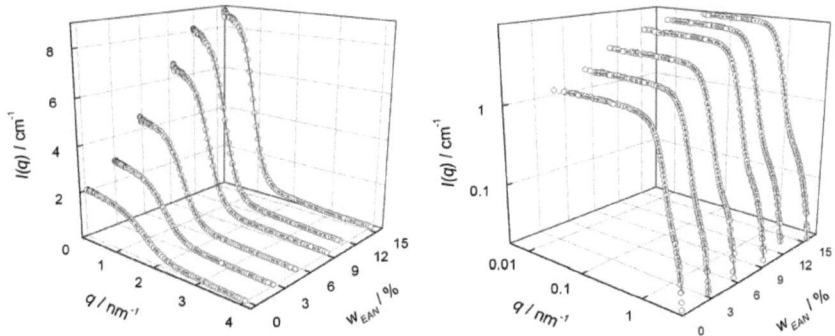

Figure IV-35. SANS curves along the experimental path for different EAN weight fractions (0 %, 3 %, 6 %, 9 %, 12 %, 16 %) at 150°C with $P_S = 40$, full lines are fit curves from the GIFT evaluation.

Both, the double linear and double logarithmic plot demonstrate the excellent fit quality from the GIFT evaluation. Furthermore, the SANS curves clearly demonstrate the high thermal stability between 30°C and 150°C along the experimental path.

The shift of the scattering peak to smaller q with increasing EAN content is also reflected in the extracted structure factor $S_{eff}(q)$ as can be seen for 30°C and 60°C in Figure IV-36 a and b and for 90°C and 150°C in Figure IV- 37 a and b, respectively.

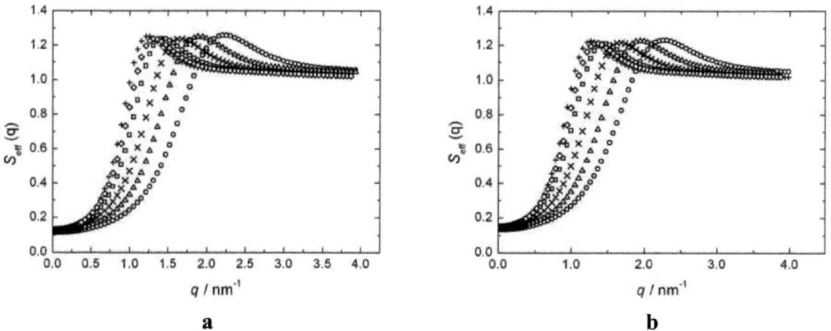

Figure IV-36. Structure factor extracted from the GIFT evaluation at 30°C (a) and 60°C (b) for different EAN weight fractions (0 % (○), 3 % (Δ), 6 % (x), 9 % (□), 12 % (◊), 16 % (+)).

IV.2.3. Ethylammonium nitrate in high temperature stable microemulsions

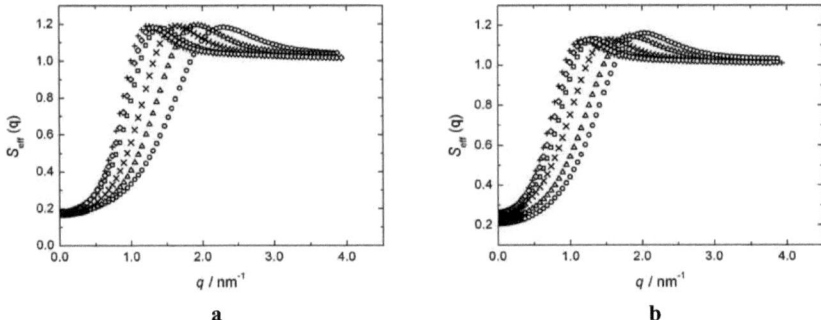

Figure IV- 37. Structure factor extracted from the GIFT at 90°C (a) and 150°C (b) evaluation for different EAN weight fractions (0 % (○), 3 % (Δ), 6 % (x), 9 % (□), 12 % (◊), 16 % (+)).

The scattering curves can be interpreted in real space in terms of their pair distance distribution function (PDDF).[178,155] This $p(r)$ function is the Fourier transform of the scattering intensity $I(q)$. We assume spherical droplets with EAN cores, the $p(r)$ function can therefore be interpreted in a classical way as histogram of distances inside the particle. The maxima of these bell shaped functions are related to the radius. The resulting pair $p(r)$ functions for 30°C and 60°C are illustrated in Figure IV-38 a and b, respectively. The pair $p(r)$ functions for 90°C and 150°C are shown in Figure IV-39 a and b, respectively.

Along the whole measured concentration range bell shaped $p(r)$ functions were obtained for all temperatures. With increasing EAN content the r_{Ai}^{max} values are shifted to the right hand side, this is in agreement with a swelling of the reverse micelles as the amount of ionic liquid increases. The r_{Ai}^{max} values are summarized in Table IV-9.

At ambient temperature, they are in the same order of magnitude compared to the R_{eff} calculated from the DLS measurements. With increasing temperature, the $p(r)$ functions become broader and the r_{Ai}^{max} increase, in particular for 150°C. Nevertheless, the increase in size with increasing temperature is not very pronounced. For the measured concentration series the shape of these functions does not change significantly. The same is true for the shape between 30°C and 150°C. This suggests that there is no essential structural variation within the measured concentration and within the measured temperature range. Nevertheless, the temperature dependent SANS experiments clearly demonstrate a thermal stability of the microemulsions between 30°C and 150°C at ambient pressure. Therefore we can assume with good confidence the existence of reverse micelles with room temperature ionic liquid cores in a continuous oil matrix over the whole investigated temperature range.

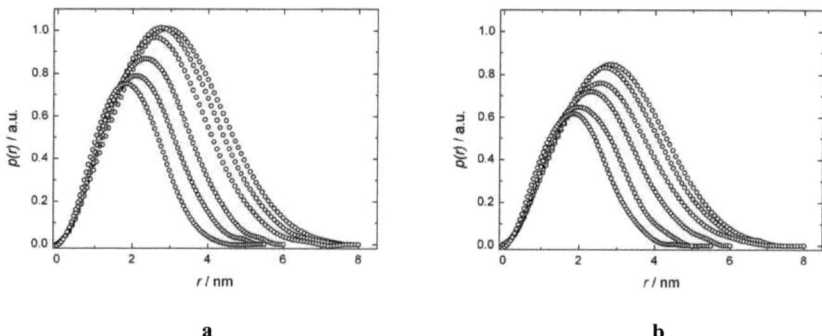

a b

Figure IV-38. PDDF functions for different EAN weight fractions (0 %, 3 %, 6 %, 9 %, 12 %, 16 %) at 30°C (a) and 60°C (b), P_S = 40, the EAN content increases from left to right.

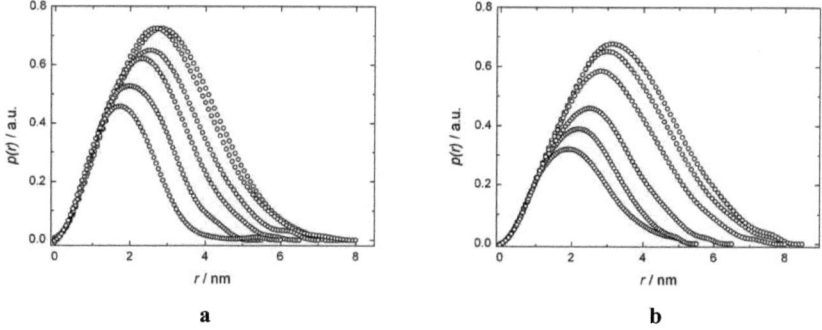

a b

Figure IV-39. PDDF functions for different EAN weight fractions (0 %, 3 %, 6 %, 9 %, 12 %, 16 %) at 90°C (a) and 150°C (b), P_S = 40, the EAN content increases from left to right.

The size of these nanodroplets can be controlled by the amount of ionic liquid. Furthermore, the change in size with increasing temperature was not very pronounced, especially between

IV.2.3. Ethylammonium nitrate in high temperature stable microemulsions

30°C and 90°C. At 150°C, the IL nanodomains were larger compared to temperatures below 100°C. Such high thermal stabilities cannot be achieved with aqueous microemulsions or microemulsions made with non-ionic surfactants which exhibit a high temperature sensitivity.

Table IV-9. r_{Ai}^{max} values obtained from the $p(r)$ functions as a function of EAN concentration and temperature.

w_{EAN} / %		0.0	2.9	5.7	9.4	11.8	15.7
r_{Ai}^{max} / nm	30°C	1.81	2.09	2.34	2.56	2.72	2.88
	60°C	1.85	1.98	2.33	2.55	2.70	2.80
	90°C	1.74	1.98	2.34	2.48	2.72	2.80
	150°C	1.92	2.20	2.54	2.80	2.98	3.15

2.3.4. Concluding remarks

In conclusion, it was demonstrated that ionic liquids in microemulsions extend the conventional thermal stability range of microemulsions at ambient pressure. We benefit in this study from the excellent thermal stability of room temperature ionic liquids to formulate nonaqueous microemulsions that are stable over a wide temperature range at ambient pressure. Using ethylammonium nitrate as polar phase, [C_{16}mim][Cl] as surfactant, decanol as cosurfactant and dodecane as continuous phase, we were able to obtain thermal stability ranging from 30°C up to 150°C.

Conductivity, dynamic light scattering and SANS measurements are in agreement with the existence of a L_2 phase (Ionic Liquid in oil). From temperature dependent conductivity measurements, a percolation phenomenon could be observed over the whole temperature range. Furthermore, the percolation threshold volume fraction was shifted with increasing temperature to lower volume fractions. DLS measurements at ambient temperature indicated a swelling of reverse micelles with increasing amount of EAN up to a certain threshold, where the formation of non-spherical structures could be assumed. This threshold is further in agreement with a percolation threshold observed via conductivity measurements. The SANS experiments were performed below this threshold. An evaluation with GIFT confirmed the existence of a L_2 phase, while the TS model that is often used to describe SAS spectra from bicontinuous microemulsions yielded unsatisfactory fit results. On the contrary, the GIFT method yields excellent fit qualities, the swelling of reverse micelles with increasing EAN content was confirmed. Furthermore, the SANS experiments clearly demonstrated the high thermal stability of the microemulsions. The shape of the pair distance distribution functions did not change significantly within the measured concentration range and within the measured temperature range. The radii of the reverse micelles increase with increasing temperature, but the effect was not very pronounced. Moreover, the *p(r)* functions can be interpreted in terms of no essential structural variation within the measured concentration and temperature range.

It should be stressed that the microemulsion chosen here was only a model system. We believe that this concept can be extended to other ionic liquids and, depending on the system, the thermal stability range can probably be much more enlarged by appropriately chosing the components. In chapter IV.2.6. high temperature stable microemulsions, where EAN was replaced by [bmim][BF$_4$] will be discussed.

These high temperature microemulsions are therefore predestined for high temperature applications such as reaction media, lubricant formulations and size controlled nanoparticle synthesis.

IV.2.4. The effect of surfactant chain length

2.4. The effect of surfactant chain length on the phase behavior of microemulsions containing EAN as polar microenvironment

2.4.1. Abstract

In this study, the chain length of the surfactant 1-alkyl-3-methyl-imidazolium chloride was varied between C_{14} and C_{18}. All other conditions are identical compared to the previous sections. Significant effects of the surfactant chain length on phase diagrams, phase behavior and thermal stability have been observed. The systems were characterized by means of dynamic light scattering, viscosity and temperature dependent conductivity measurements.

2.4.2. Results and discussion

2.4.2.1. Phase diagrams

The phase diagrams at 30°C have been determined according to the procedure described in section IV.2.3.3.1. The resulting pseudo-ternary phase diagrams, where the surfactant chain lengths vary between C_{14}, C_{16} and C_{18} are illustrated in Figure IV-40 a, b and c, respectively.

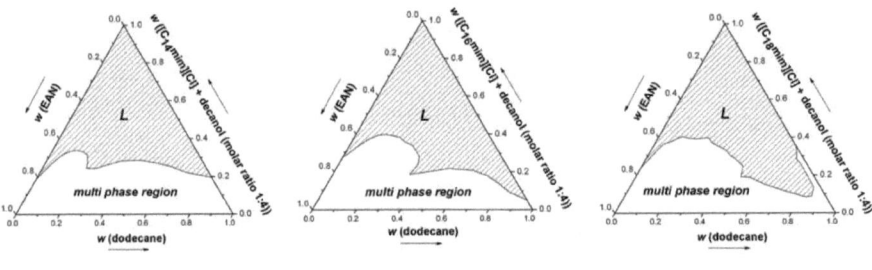

Figure IV-40. Pseudo ternary phase diagrams of dodecane/([$C_n$mim][Cl]+decanol)/ EAN with n = 14 (a), 16 (b) and 18 (c). L represents the clear and isotropic single phase region.

For the three phase diagrams, a huge clear and isotropic single phase region (L) was observed. The topologies of the phase diagrams show similarities, but also some systematic differences. For the C_{14} and C_{16} surfactants, the phase diagrams exhibit an appendage towards 100 % EAN indicating a partial nonionic character of the surfactant headgroup. For all phase diagrams the phase boundaries are irregular. This effect becomes more pronounced with increasing surfactant chain length. The unmixing limit of EAN and the pseudo-constituents with C_{14}, C_{16} and C_{18} chain lengths were 81 % (w/w), 72 % (w/w), 76 % (w/w), respectively. Probably the most important difference between the phase diagrams can be seen along the surfactant+cosurfactant axes. For a C_{14} surfactant chain length, the unmixing limit is below 20 wt% of the pseudo-constituent. With increasing chain length (C_{16}), the solubility of the pseudo-constituent increases. This observation can be related to the increasing hydrophobicity with increasing chain length, the unmixing limit is depressed down to below 5 wt%. When the surfactant chain length is further increased the unmixing limit increases again, 35 wt% of the pseudo-constituent are necessary to obtain a clear and homogeneous solution. This finding appears at first view surprising, as the hydrophobicity of the surfactant increases. On the other hand the structure of microemulsions plays an important role on the solubization. For the C_{18} system, the formation of liquid crystals was observed at low EAN contents. Hence, one possible explanation is that with increasing chain length the order of the formed structures increases. The interfacial film becomes less flexible with increasing chain length. Thus, a balance between increasing hydrophobicity and rigidity of the film can be assumed. The appendage observed for the C_{14} and C_{16} surfactant towards 100 % EAN becomes less pronounces which is a further hint for a less flexible interfacial film for the C_{18} surfactant.

By taking the same experimental path with $P_S = 40$ as in the previous chapter, no phase change could be observed visually between 30°C and 150°C.

2.4.2.2. Density

For the calculation of the droplet volume fraction densities, ρ, of the pseudo constituent were measured with a densimeter DMA60A between 25°C and 55°C. It was abstained from using a pyknometer to cover the whole temperature range between 25°C and 150°C as a linear dependence of the density as a function of temperature was assumed and otherwise 20 mL in volume would have been necessary for each measurement. This assumption appears to be justified as density measurements between 25°C and 150°C for [C_{16}mim][Cl]+decanol yielded a linear density temperature relationship over the whole temperature range (compare IV.2.3.3.1). The following linear density–temperature relationships were obtained:

$\rho_{[C18mim]Cl/1\text{-}Decanol}$ / g cm^{-3} = 0.879 – 0.0004 θ /°C

$\rho_{[C14mim]Cl/1\text{-}Decanol}$ / g cm^{-3} = 0.880 – 0.0004 θ /°C

IV.2.4. The effect of surfactant chain length

2.4.2.3. Conductivity

Conductivities were measured along the same experimental path used for the investigations with the C_{16} chain length with $P_S = 40$ and variable EAN content. Specific conductivities for the C_{14} and C_{18} chain length are illustrated in Figure IV-41 and Figure IV-42, repectively. Both systems exhibit a clear percolation phenomenon similar to the investigation with [C_{16}mim][Cl] described in the previous section. Merely at 150°C of the C_{14} microemulsion, the shape of the curve looks different. This gives hints concerning either a structural difference or a reduced thermal stability with decreasing surfactant chain length. Such speculations are confirmed by regarding the phase behavior with a [C_{12}mim][Cl] as surfactant. A long the same experimental path a phase separation occurred at elevated temperatures ($\theta > 120°C$). For this reason, it was refrained from high temperature measurements with the C_{12} chain length.

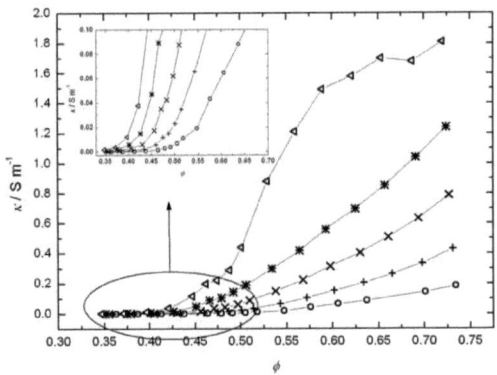

Figure IV-41. Specific conductivity for the C_{14} chain length as a function of the droplet volume fraction ϕ demonstrating the percolative behavior at different temperatures (30°C (○), 60°C (+), 90°C (×), 120°C (∗) and 150°C(Δ)).

The tendency observed for microemulsions with [C_{16}mim][Cl] that the percolation threshold is shifted with increasing temperature to lower volume fractions was confirmed for the C_{14} and C_{18} chain length. The corresponding percolation threshold volume fractions ϕ_P were determined from the inflection point of the curves log (κ) = f (ϕ). The curves with a fourth order polynomial fit (full lines) are shown in Figure IV-43 for microemulsion with

[C$_{14}$mim][Cl] and Figure IV-44 for [C$_{18}$mim][Cl], respectively.

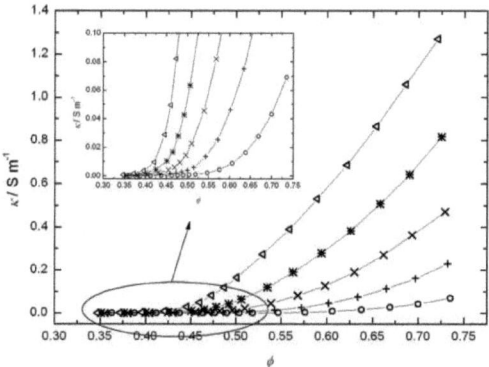

Figure IV-42. Specific conductivity for the C$_{18}$ chain length as a function of the droplet volume fraction ϕ demonstrating the percolative behavior at different temperatures (30°C (○), 60°C (+), 90°C (×), 120°C (∗) and 150°C(Δ)).

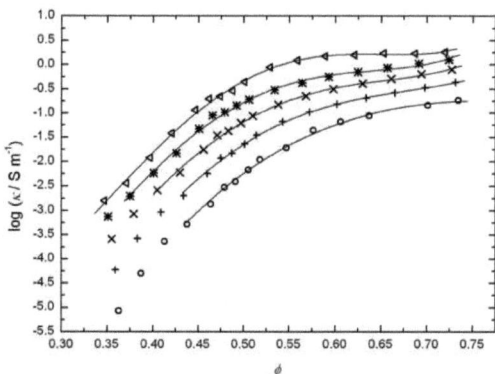

Figure IV-43. Determination of the percolation threshold volume fraction for the C$_{14}$ chain length at various temperatures (30°C (○), 60°C (+), 90°C (×), 120°C (∗) and 150°C(Δ)), full lines represent fourth order polynomial fits.

IV.2.4. The effect of surfactant chain length

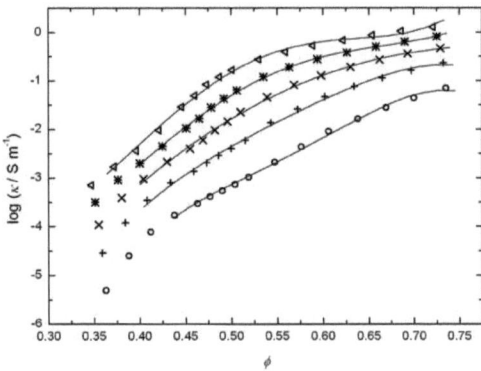

Figure IV-44. Determination of the percolation threshold volume fraction for the C_{18} chain length at various temperatures (30°C (○), 60°C (+), 90°C (×), 120°C (*) and 150°C(Δ)), full lines represent fourth order polynomial fits.

The resulting percolation threshold volume fractions in dependence of temperature and surfactant alkyl chain length n are summarized in Table IV-10. For the C_{14} and C_{16} chain length, ϕ_P values are very similar, for the C_{18} chain, ϕ_P values between 30°C and 90°C are significantly higher. These findings indicate again that the interfacial film in the case of the C_{18} surfactant is less flexible. Hence, the threshold is shifted to higher volume fractions in the low temperature region. Obviously, this effect disappears at higher temperatures. For all systems ϕ_P depends almost linearly on temperature.

Table IV-10. Temperature dependence of the percolation threshold volume fraction for different surfactant chain lengths.

θ / °C	30°C	60°C	90°C	120°C	150°C
			ϕ_P		
C_{14}	0.49	0.46	0.43	0.40	0.37
C_{16}	0.48	0.46	0.43	0.41	0.39
C_{18}	0.54	0.50	0.47	0.43	0.39

2.4.2.4. Viscosity

The dynamic viscosities at 30°C of the microemulsions exhibited Newtonian behavior along the whole investigated experimental path. The dynamic viscosities for the C_{14} and C_{18} chain lengths versus EAN volume fractions are visualized in Figure IV-45 a and b, respectively.

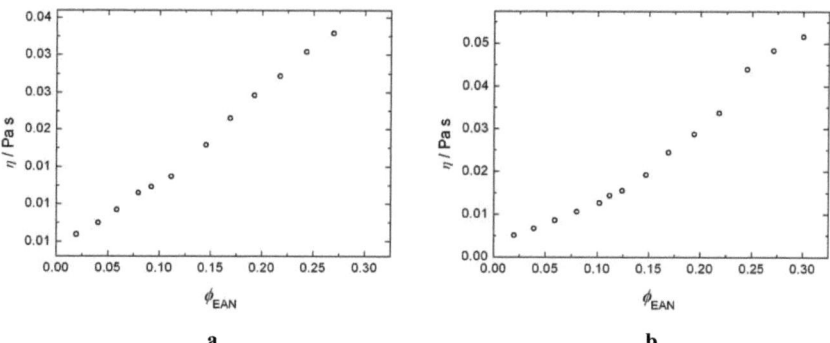

a b

Figure IV-45. Dynamic viscosities at 30°C in dependence of the volume fraction of EAN for the C_{14} (a) and the C_{18} (b) chain length.

The viscosities increase with increasing amount of EAN. Compared to the conductivities, the viscosities did not increase over several orders of magnitude. Therefore, it was not possible to confirm the percolation behavior of the EAN system. One condition to observe a viscosity percolation phenomenon in microemulsions is that the viscosity of the continuous phase and the polar phase are significantly different. This condition is not fulfilled for the present systems. However, the measurements clearly show the relatively low viscosity of the present microemulsions. The viscosities of microemulsions formed with a C_{14} and a C_{16} chain length were very similar. On the contrary, the viscosity of the microemulsions with the C_{18} chain length was very similar for low EAN weight fractions (< 15 wt%) and increased above this value more pronounced.

2.4.2.5. Dynamic light scattering

Similar to the C_{16} system, a monomodal decay in the autocorrelation function was observed for the C_{14} and C_{18} systems for EAN weight fractions below the percolation threshold. The corresponding apparent hydrodynamic radii as a function of EAN weight fraction are given in

IV.2.4. The effect of surfactant chain length

Figure IV-46. At higher EAN amounts a bimodal decay was observed indicating the formation of nonspherical structures or the formation of droplet clusters. The order of magnitude of R_{Happ} at a given EAN content is comparable for all three systems. No significant change in size with increasing surfactant chain length could be detected with DLS. This may also be related to the scattering contrast mainly given by EAN and the surfactant head group. However, DLS measurements indicate droplet structures below the percolation threshold.

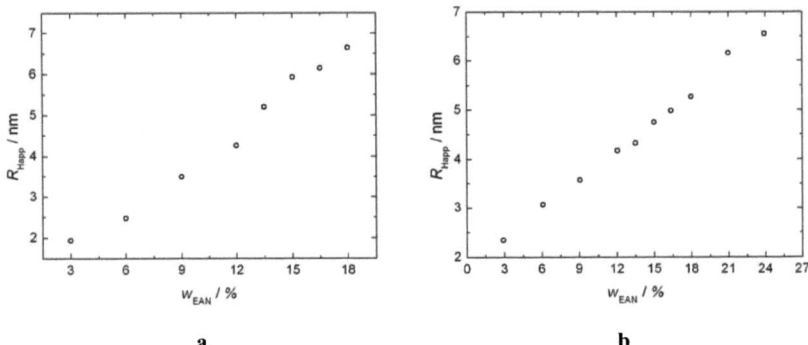

a b

Figure IV-46. Apparent hydrodynamic radii for those EAN weight fractions where single exponential decay was observed indicating a swelling of the reverse micelles with increasing EAN content at 30°C for the C_{14} (a) and the C_{18} (b) chain length.

2.4.3. Concluding remarks

In conclusion, it was demonstrated that the surfactant chain length plays an important role on the phase diagram topology, phase behavior and flexibility of the interfacial film. To obtain comparable systems, all conditions, except the surfactant chain length were kept as in the previous sections. From the resulting pseudo-ternary phase diagram, different topologies of the single phase region have been found. This was related to two different effects, on the one hand, the solubility of the surfactant-cosurfactant mixture in the oil phase increases with increasing chain length. Consequently, the area of the one phase region increased from C_{14} to C_{18}. On the other hand, the structural order increased with increasing surfactant chain length. This was mainly indicated by the formation of liquid crystals for the C_{18} chain length at low surfactant weight fraction. It was assumed that the flexibility of the surfactant film decreases with increasing chain length. Consequently, the amount of surfactant-cosurfactant mixtures necessary for the formation of a clear, isotropic homogeneous solution with dodecane increases from C_{16} to C_{18}. For all system, irregular phase boundaries have been found. A single phase appendage towards 100 wt% EAN was observed for all systems being less pronounced for the C_{18} chain length. Conductivity measurements demonstrated a percolative behavior for all three microemulsion systems. For the C_{14} and C_{16} chain length, the temperature dependent percolation threshold volume fractions were very similar at a given temperature. For the C_{18} chain length, ϕ_P was significantly higher at 30°C, 60°C and 90°C. These findings underline the increasing rigidity of the interfacial film for microemulsions with the C_{18} chain length. Viscosity measurements did not show any percolation effect for all three systems. Viscosities were in the same order of magnitude at a given EAN content for a C_{14} and C_{16} surfactant chain length, whereas they were slightly higher for microemulsions with [C_{18}mim][Cl]. Dynamic light scattering experiments confirmed a swelling behavior with increasing amount of EAN for samples with EAN weight fractions below the percolation threshold. Above this threshold a bimodal decay in the intensity autocorrelation function was observed indicating the formation of non-spherical structures or droplet clusters. No significant effect in size with increasing chain length from the DLS measurements could be reported. The effect of the oil continuous phase remains to be discussed.

2.5. Biodiesel, a sustainable oil, in high temperature stable microemulsions containing a low-toxic room temperature ionic liquid as polar phase

2.5.1. Abstract

In the previous section, the effect of the surfactant chain length has been discussed. In the following, the oil phase was varied, dodecane was replaced by biodiesel.

Biodiesel has gained more and more attention in recent years resulting from the fact that it is made from renewable resources. Parallel to its environmental compatibility, biodiesel also exhibits a high thermal stability. We demonstrate here that biodiesel can replace conventional oils as apolar phase in nonaqueous microemulsions containing the room temperature ionic liquid ethylammonium nitrate as polar phase. In addition to the phase diagram and the viscosity of the microemulsions, we study the thermal stability of these systems. Along an experimental path in the phase diagram, no phase change in the microemulsion structure could be observed between 30°C and 150°C. Conductivity measurements confirm the high thermal stability of these systems. The microemulsion exhibit a percolative behavior between 30°C and 150°C. Small angle X-ray scattering spectra show a single broad scattering peak similar to aqueous microemulsions. The spectra could well be described by the Teubner-Strey model. Furthermore, the adaptability of different models ranging from bicontinuous structures to ionic liquid in oil spheres has been checked. These high temperature stable, nonaqueous, free of crude oil based organic solvent microemulsions highlight an efficient way towards the formulation of environmentally compatible microemulsions and open a wide field of potential applications.

2.5.2. Introduction

Sustainable and renewable energy sources have become more attractive due to their environmental benefits. In this context, biodiesel, basically used as alternative diesel fuel, is made from renewable energy sources, it is biodegradable and nontoxic.[1] Its environmental compatibility also makes it interesting to replace conventional oils in microemulsions. The formation of aqueous microemulsions with vegetable oils and diesel has already been described in literature.[254,255] Hazbun *et al.* reported water/fuel microemulsions with diesel fuel or kerosene as oil phase with a thermal stability ranging from -10°C to 70°C.[256] Biodiesel fuels mainly consist of methylesters of renewable oils, fats and fatty acids.[257,258,259] The potential of biodiesel as new green solvent has also been described in literature.[260] Recently, Wellert *et al.* presented aqueous biodiesel based microemulsions as media for environmentally friendly decontamination.[261] Parallel to its environmental compatibility, biodiesel also exhibits a high thermal stability.

Studies concerning ILs in microemulsions, especially EAN has been discussed in detail in the previous chapters.

In this chapter, we combine both, biodiesel as oil phase and room temperature ionic liquid, namely ethylammonium nitrate as polar phase. Further we use the long chain ionic liquid 1-hexadecyl-3-methylimidazolium chloride as surfactant and decanol as cosurfactant. All ingredients show an excellent thermal stability making these systems interesting for high temperature applications. The microemulsions formed with biodiesel as oil phase are compared to microemulsions formed with dodecane as oil phase. Similarities and differences of these to systems are highlighted.

2.5.3. Results and discussion

2.5.3.1. Phase diagram

For the pseudo-ternary phase diagram the surfactant ($[C_{16}mimCl]$)/cosurfactant (decanol) molar ratio was kept constant at 1:4 as described for the previous experiments. The resulting phase diagram is illustrated in Figure IV-47 a, where L represents the clear and homogenous single phase region. All ingredients show an excellent thermal stability that is necessary for temperature dependent investigations. A huge single phase region could be observed. The topology of the phase diagram exhibit many similarities compared to the phase diagram obtained when biodiesel is replaced by dodecane as can be seen from Figure IV-47 b.

The single phase nose towards 100 wt % EAN observed for the dodecane system is less pronounced for biodiesel. Furthermore, the single phase region at low surfactant contents was larger in the case of biodiesel. This observation is related to solubility of the surfactant-cosurfactant mixture in the apolar phase, which is increased in biodiesel compared to dodecane. As biodiesel consists mainly of fatty acid methyl esters, the polarity of these substances is higher compared to a linear hydrocarbon. Hence, the solubility of the surfactant-cosurfactant mixture increases in biodiesel. The area of the single phase region is comparable for both systems. In order to compare the biodiesel and dodecane microemulsions, the same experimental path, where the amount of surfactant plus cosurfactant was kept constant at a

IV.2.5. Biodiesel in high temperature stable microemulsions

weight fraction of 40 % (P_S = 40) was chosen for all the following investigations.

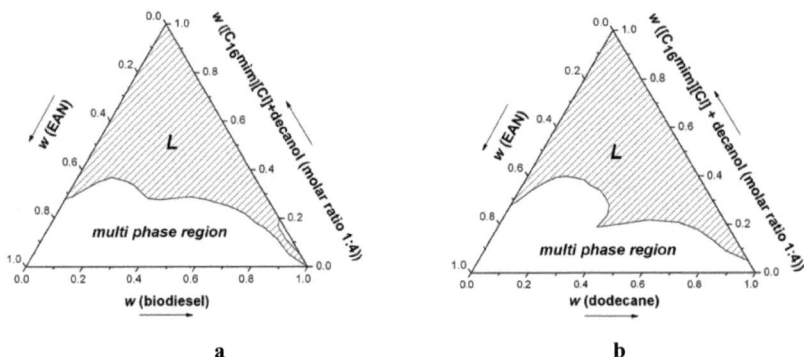

a b

Figure IV-47. Pseudo-ternary phase diagram at 30°C for biodiesel a) and dodecane b), respectively.

2.5.3.2. Visual observations

Along the experimental path, no phase change could be observed visually within a temperature range between 30°C and 150°C. Photos of the microemulsion at 30°C and 150°C, respectively indicating the excellent thermal stability of the systems are shown exemplarily for w_{EAN} = 12 % in Figure IV-48.

Figure IV-48. Photos of the microemulsions for P_S = 40 and w_{EAN} = 12 % at 30°C and 150°C, respectively.

All solutions were stable for a time period of at least three months. Afterwards, the former light yellow solutions turned dark yellow. This observation can be related to a partial degradation or oxidation of the biodiesel phase. Hence, all solutions for the following

2.5.3.3. Density

Densities, ρ, of biodiesel were measured with a pyknometer within a temperature range of 25°C ≤ θ / °C ≤ 150°C ± 0.1°C in intervals of 10°C. The temperature dependent densities of the surfactant-cosurfactant mixture have already been reported in section IV.2.3.3.1. Densities of biodiesel are shown in Figure IV-49, a linear temperature density relationship could be observed yielding:

$$\rho_{biodiesel} / \text{g cm}^{-3} = 0.894 - 0.0007 \, \theta \, / \, °C$$

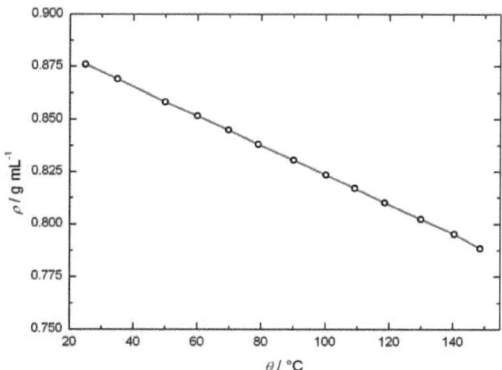

Figure IV-49. Densities, ρ, of biodiesel within a temperature range between 25°C and 150°C, the full line represents a linear fit.

The densities were used for the calculation of the volume fraction of the dispersed phase, ϕ, which was calculated from the sample composition with the assumption of ideal mixing.

2.5.3.4. Conductivity

As already discussed in detail in the previous sections, conductivity measurements give important information about the percolative or antipercolative behavior of microemulsions. The percolation phenomenon can be observed, if the specific conductivities $\kappa_{polar} \gg \kappa_{oil}$. As the difference in conductivity between EAN and biodiesel (oil) is very large, this condition is fulfilled. For the present study, the conductivity was measured along the above mentioned

IV.2.5. Biodiesel in high temperature stable microemulsions

experimental path at different temperatures ranging from 30°C to 150°C in intervals of 30°C. Figure IV-50 a and b show the specific conductivity along the experimental path at 30°C and 150°C, respectively. It can be seen, that the conductivity at a constant amount of EAN increases with increasing temperature. Furthermore, the conductivity indeed makes a sharp increase over several orders of magnitude with increasing amount of EAN.

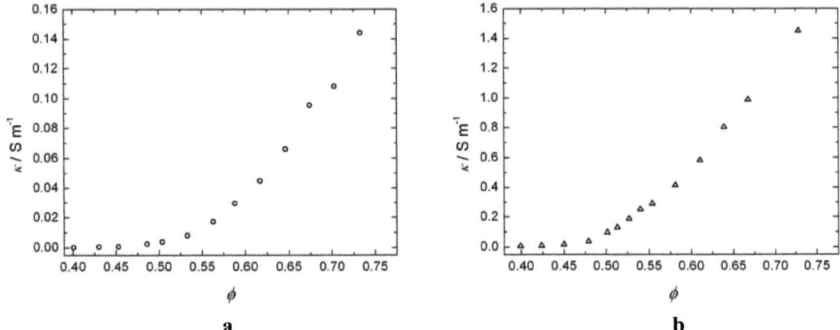

Figure IV-50. Specific conductivity as a function of the volume fraction ϕ at 30°C (a) and 150°C (b), respectively.

For all measured temperatures a percolative behavior could be observed. This is illustrated in Figure IV-51. From the inset of the plot it can be seen that the shift of the percolation threshold volume fraction is less pronounced and very weak compared to the dodecane system. The corresponding percolation threshold volume fraction ϕ_P was calculated from the inflection point of the curve $log(\kappa) = f(\phi)$ as shown in Figure IV-52. These curves exhibit a sigmoid shape. $log(\kappa)$ was fitted to a fourth order polynomial in ϕ from which the inflection point was determined in order to obtain ϕ_P.

IV. Results and discussion

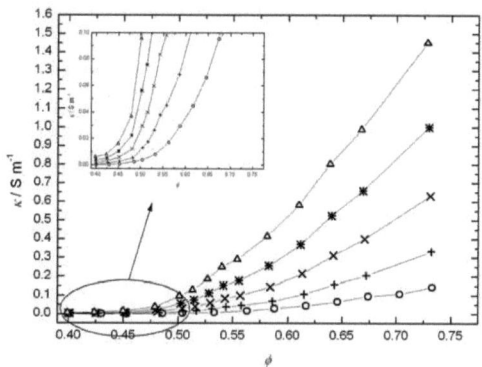

Figure IV-51. Specific conductivity as a function of the volume fraction ϕ demonstrating the percolative behavior at different temperatures (30°C (○), 60°C (+), 90°C (×), 120°C (*) and 150°C(Δ)).

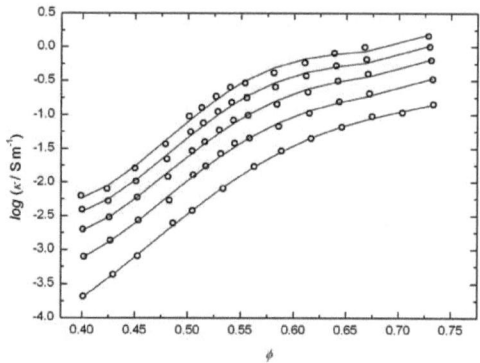

Figure IV-52. Determination of the percolation threshold volume fraction, full lines represent fits with a fourth order polynomial.

The resulting temperature dependent percolation threshold volume fractions ϕ_P are summarized in Table IV-11. On the contrary to similar microemulsions formulated with dodecane, ϕ_P seems to be nearly independent of temperature. Nevertheless, the conductivity measurements confirm the visually observed high thermal stability of the microemulsions.

Table IV-11. Temperature dependence of the percolation threshold volume fraction ϕ_P.

θ / °C	30	60	90	120	150
ϕ_P	0.47	0.48	0.48	0.48	0.48

These findings are surprising as a quite strong temperature dependence concerning microemulsions percolation was expected. These results underline important differences concerning the nature of the oil phase in microemulsions. To obtain more information about shape and size of the microemulsions dynamic light scattering experiments have been performed. Unfortunately, these experiments did not yield an intensity autocorrelation function that could be used for a convenient data evaluation. Hence, SAXS experiments have been performed in order to prove a true microemulsion structure.

2.5.3.5. Small angle X-Ray scattering (SAXS)

In order to get more inside into the microemulsions structure, particularly about the shape and size of the structures SAXS experiments have been performed at 30°C. The curves exhibit a single broad scattering peak as it was observed with SANS experiments for microemulsions with dodecane as continuous phase. At higher q values, a characteristic q^{-4} dependence was observed. The scattering intensity increases and the peak maxima are shifted to smaller q values with increasing EAN weight fraction. The TS model was used to analyze the scattering curves. The SAXS curves in double linear and double logarithmic plot are shown in Figure IV-53 a and b, respectively.

From the fit, the domain size, d, the correlation length, ξ, and the amphiphilic factor, f_a, were extracted. These parameters are summarized in Table IV-12. The domain size of the microemulsions increases with increasing amount of EAN which can be correlated to a swelling of the formed structures. The correlation length remains constant within the uncertainty limits. The values for the amphiphilic factor are in the region of well-structured microemulsions. A significant increase in the amphiphilic factor with increasing amount of EAN can be reported.

IV. Results and discussion

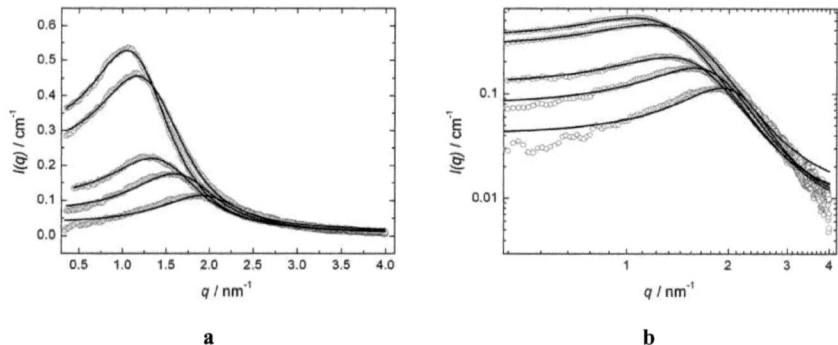

a   b

Figure IV-53. SAXS curves along the experimental path for different EAN weight fractions (3 %, 6 %, 9 %, 12 %, 16 %) in the double linear view (a) and double logarithmic view (b), full lines fit with the TS formula, the amount of EAN increases from the bottom up.

Table IV-12. Characteristic length scales and amphiphilic factor from the TS fit.

w_{EAN} / %	d / nm	ξ / nm	f_a
3.1	3.1	1.6	-0.83
6.0	3.7	1.6	-0.76
9.2	4.2	1.5	-0.68
12.0	4.7	1.6	-0.63
15.7	5.2	1.7	-0.61

The specific area Σ at the polar/non-polar interface can be obtained from the SAXS spectra in the large q-range where the scattered intensity follows a q^{-4} behavior, this so-called Porod regime was calculated according to eq. IV-10.[247] This regime is only obtained, if a thin interface separates two media of different scattering length densities. For the present systems a Porod limit could indeed be observed at large q values after the background subtraction. The Porod limit at large q is shown exemplarily for 9 wt % EAN in Figure IV-54.

IV.2.5. Biodiesel in high temperature stable microemulsions

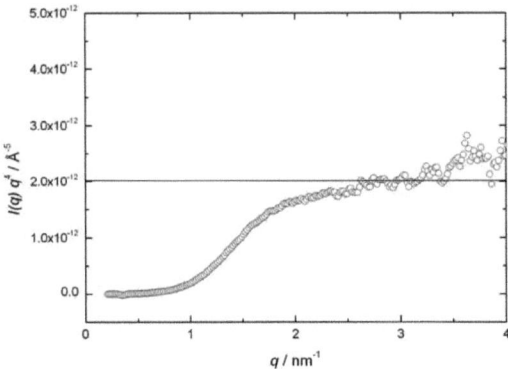

Figure IV-54. Porod limit shown exemplarily for 9 wt % EAN, the straight line denotes the Porod limit.

A well-known problem concerning the evaluation of $\Delta\rho$ is the distribution of decanol in dodecane and the interfacial film, respectively. Decanol is partially soluble in dodecane and insoluble in EAN. To obtain convenient upper and lower limits for the specific surface Σ, a parameter α was defined that takes the distribution of decanol in the interfacial film and the oil phase into account. For $\alpha = 1$, the OH groups of decanol are quantitatively located in the interfacial film. For $\alpha = 0$, all decanol is in the alkane phase and does not contribute to the interfacial film. As Biodiesel is mainly a mixture of different fatty acid methyl esters, several assumptions had to be made. The scattering length density is an estimation to determine sign and order of magnitude of the value. Neglecting all components except the methyl ester of the C_{18} fatty acid a scattering length density $\rho_{Biodiesel} = 8.44 \; 10^{10} \; cm^{-2}$ was obtained using a mean molar mass of 285 g mol^{-1} and a density of 0.873 g cm^{-3}. The results for the Porod limit, the invariant and the specific surface for different α values between 0 and 1 are summarized in Table IV-13.

With increasing α, Σ increases for a constant volume fraction of RTIL. Moreover, Σ increases with the amount of EAN. For the interpretation of the result the adaptability of several models was checked, namely the cubic random cell model,[172] spheres of IL/o[173] and repulsive spheres.[157] For the EAN system, the experimental curve lies between the two sphere models confirming the existence of spherical structures. Interestingly, these results are comparable to microemulsions with dodecane as oil phase as it was shown in section IV.2.2. The existence of spherical structures with EAN cores is in agreement with the conductivity data as well as the SAXS curves measured for EAN weight fractions below the percolation threshold.

Table IV-13. Σ for the EAN microemulsions along the experimental path for different α values with the corresponding invariant and Porod limit.

ϕ_{EAN} / %		2.27	4.42	6.83	9.04	11.96
$\phi_{biodiesel}$ / %		57.77	55.24	52.51	49.91	44.96
$\phi_{decanol}$ / %		26.71	26.96	27.18	27.44	28.80
$\phi_{[C16mim][Cl]}$ / %		13.25	13.37	13.48	13.61	14.28
Q_{exp} / (cm^{-4}/10^{21})		1.47	1.52	1.44	2.06	1.89
Porod limit / (cm^{-5}/10^{27})		25.0	23.0	20.0	24.0	19.0
Σ / (cm^2/cm^3/10^6)	α = 0.00	3.41	3.90	4.42	4.33	4.43
	α = 0.25	3.73	4.18	4.66	4.52	4.59
	α = 0.50	4.06	4.45	4.90	4.71	4.74
	α = 0.75	4.37	4.72	5.13	4.89	4.89
	α = 1.00	4.68	4.99	5.36	5.07	5.04

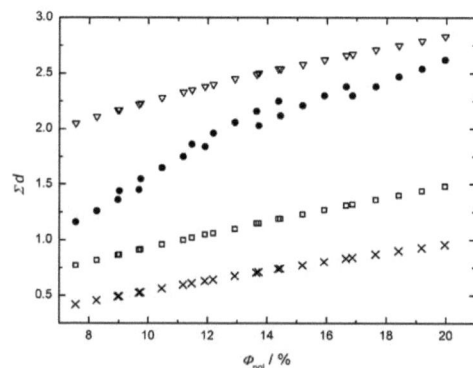

Figure IV-55. Σd versus Φ_{pol} for the EAN system of the experimental data ●, the cubic random cell model x, the model of w/o spheres ▲ and the model of repulsive spheres ☐ for α between 0 and 1 (0.00; 0.25; 0.50; 0.75; 1.00).

2.5.3.6. Viscosity

Viscosity measurements are an important tool for the characterization of microemulsions. Similar to conductivity measurement, equations for percolation have been suggested,[135] but a quantitative verification is only possible, if the viscosity η of the polar compound is sufficiently different from η of the oil.[74] Along the experimental path all samples exhibited Newtonian behavior within the measured shear rates. The mean value of the dynamic viscosities increases slightly with increasing amount of room temperature ionic liquid as shown in Figure IV-56. They lie in the same order of magnitude as observed for similar microemulsions where biodiesel is replaced by dodecane.[195] Nevertheless, no viscosity percolation effect could be detected.

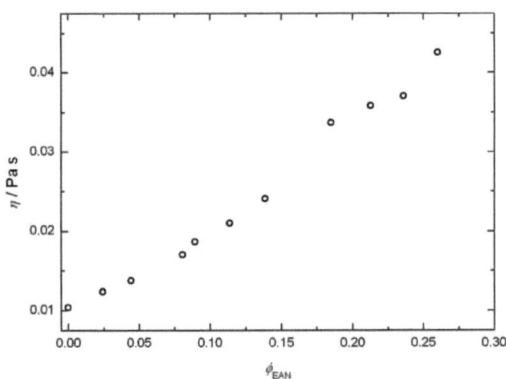

Figure IV-56. Mean dynamic viscosities along the experimental path at 30°C.

2.5.4. Concluding remarks

Over the last years, environmental compatible solvents have attracted more and more attention. We demonstrate here that biodiesel can replace alkanes in non-aqueous ionic liquid containing microemulsions. The phase diagram observed with biodiesel exhibits a huge clear and isotropic single phase region. Due to the excellent thermal stability of biodiesel, the surfactant, cosurfactant and the room temperature ionic liquid ethylammonium nitrate, the thermal stability along an experimental path was studied visually and with temperature dependent conductivity measurements between 30°C and 150°C. These experiments clearly demonstrate the pronounced thermal stability of these microemulsions at atmospheric pressure. Furthermore, the viscosity of the microemulsions was comparable to the viscosity when biodiesel was replaced by dodecane. Conductivity measurements indicated percolation phenomena when the amount of EAN was increased. Contrary to similar microemulsions with dodecane, the percolation threshold volume fraction seems to be relatively insensitive to temperature and does not change significantly with temperature. SAXS experiments demonstrated the formation of microemulsions, a single broad scattering peak was observed followed by a characteristic q^{-4} dependence at large q. The curves could well be described with the TS formula. From the specific surface three different models ranging from bicontinuous structures to IL/o droplets have been checked. Below the percolation threshold the formation of reverse micelles with EAN cores was confirmed. Microemulsions made of sustainable oil provide a natural way to "greener" formulations. Furthermore, the cytotoxicity of EAN is significantly lower compared to imidazolium based ionic liquids.[262] The next step is, evidently, to substitute low-toxic and biodegradable surfactants for imidazolium based surfactants. These microemulsions containing room temperature ionic liquids as polar phase exhibit a thermal stability that cannot be obtained with aqueous microemulsions at atmospheric pressure. Such extensions of the conventional thermal stability range of microemulsions open a wide field of potential applications such as extractions, organic synthesis and nanoparticles synthesis.

2.6. [bmim][BF$_4$] in high temperature stable microemulsions

2.6.1. Introduction

In section IV.2.2 microemulsions with EAN and [bmim][BF$_4$] as polar microenvironment have been compared at ambient temperature. It was demonstrated that the nature of the RTIL plays an important role on phase diagrams, phase behavior and microstructure of the microemulsions. Compared to the EAN system, a much higher amount of surfactant+cosurfantant was necessary to obtain microemulsions with [bmim][BF$_4$]. Due to the high amount of surfactant used, the obtained microstructure at ambient temperature was assumed to be more a bicontinuous one than a L_2 phase microemulsion in contrary to the EAN system. Neither with conductivity nor with viscosity measurements any percolation phenomenon could be observed. In this chapter the stability at high temperatures of these microemulsions is investigated with temperature dependent conductivity and temperature dependent SANS experiments.

2.6.2. Results and discussion

2.6.2.1. Conductivity

The electrical conductivity of the microemulsions has been studied for the same experimental path used for the investigations at ambient temperature with a constant weight fraction of surfactant+cosurfactant ($P_S = 65$) and increasing amount of RTIL. The measured temperature range varied between 30°C and 150°C in temperature intervals of 30°C. Comparable to the investigations at ambient temperature, a continuous increase in conductivity with increasing [bmim][BF$_4$] weight fraction has been observed, no sharp threshold could be detected. Furthermore, the conductivity increases with increasing temperature for a constant [bmim][BF$_4$] weight fraction. The conductance at Ps= 65 without the addition of RTIL was significantly higher compared to P_S= 40. One possible explanation is that the high amount of surface active agent does not allow the formation of well separated structures with oil as continuous phase. For the calculation of the [bmim][BF$_4$] volume fraction, temperature dependent densities were necessary. Temperature dependent densities and conductivities of [bmim][BF$_4$] have been reported by Stoppa *et al* between 5°C to 65°C in intervals of 10°C.[263] The density data could well be described by a linear fit. The following linear density temperature dependence was used to extrapolate densities up to 150°C.

$\rho_{[bmim][BF4]}$ / g cm^{-3} = 1.219 − 0.0007 θ / °C

In Figure IV-57 the specific conductivities as a function of the [bmim][BF$_4$] volume fraction, $\phi_{[bmim][BF4]}$ are shown. From the inset of the plot one can see a continuous increase in conductivity with increasing amount of [bmim][BF$_4$] between 30°C and 150°C, no percolation threshold could be detected.

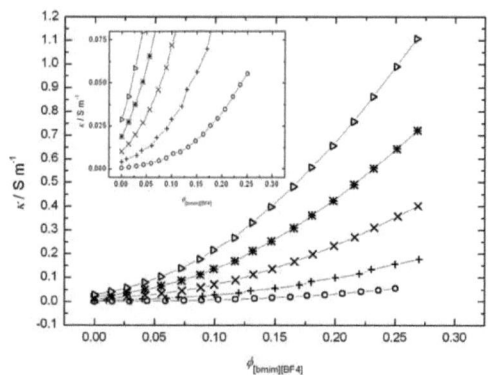

Figure IV-57. Specific conductivity as a function of the RTIL volume fraction $\phi_{[bmim][BF4]}$ at different temperatures (30°C (○), 60°C (+), 90°C (×), 120°C (*) and 150°C(Δ)).

Nevertheless, it is of interest to compare the conductivities to those of pure [bmim][BF$_4$] to ensure that dodecane is still the continuous phase. We recently reported precise temperature dependent conductivity data of [bmim][BF$_4$] within a temperature ranging from -35°C to 195°C.[263] The temperature dependent conductivities are illustrated in Figure IV-58 in comparison to data available in literature, the full line represent a VFT fit. To compare these data of pure [bmim][BF$_4$] to the microemulsions three representative points in the phase diagram along the experimental path have been chosen: one at low [bmim][BF$_4$] content ($\phi_{[bmim][BF4]}$ = 3 %), one in the intermediate region ($\phi_{[bmim][BF4]}$ = 15 %) and one point with a high RTIL content ($\phi_{[bmim][BF4]}$ = 25 %). These points were compared to pure [bmim][BF$_4$] at 5 different temperatures (30°C, 60°C, 90°C, 120°C, 150°C) as shown in Figure IV-59. From this graph, two important conclusions can be drawn out. The molar conductivity at each composition of the microemulsions is significantly lower compared to the neat RTIL

IV.2.6. [bmim][BF$_4$] in high temperature stable microemulsions

indicating that [bmim][BF$_4$] cannot be considered as continuous phase.

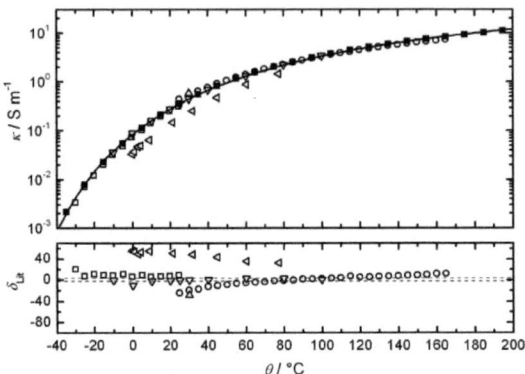

Figure IV-58. Conductivities, κ, of [bmim][BF$_4$] from Stoppa *et al.*[263] (■) and relative deviations, δ_{Lit}, of appropriately interpolated values from published data of Suarez *et al.*[211] (◁), Nishida *et al.*[264] (□), Tokuda *et al.*[214] (∇), Liu *et al.*[265] (Δ) and Villa *et al.*[266] (○). The solid line represents a VFT fit, dashed lines indicate a (arbitrary) $\delta_{Lit} = \pm 3$ margin.

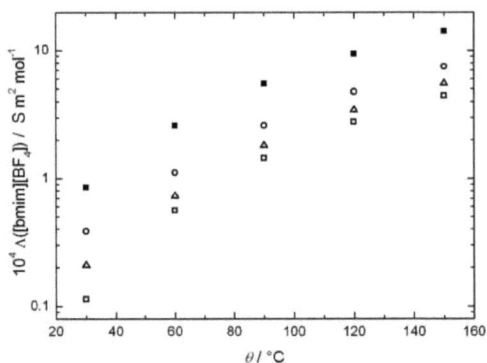

Figure IV-59. Comparison of the molar conductivities for neat [bmim][BF$_4$] (■) of appropriately interpolated values from Stoppa *et al.*[263] and microemulsions with $\phi_{[bmim][BF4]} = 3\ \%$ (○), 15 % (Δ) and 25 % (□) as a function of temperature.

Furthermore, for a given microemulsion composition the difference in molar conductivity, Δ, compared to neat [bmim][BF$_4$] does not change with increasing temperature. This fact can be interpreted assuming that there is no essential structural variation as a function of temperature. Moreover, conductivity measurements confirm the high thermal stability of the microemulsions and indicate that the formed structures are not well separated from each other. Bicontinuous instead of droplet like structure can be assumed.

2.6.2.2. Small angle neutron scattering (SANS)

To obtain more information about shape and size and thermal stability of the microemulsions temperature dependent small angle neutron scattering experiments (SANS) have been performed. An evaluation with the GIFT method was not possible in this case because of the high amount of surfactant used ($P_S = 65$). As conductivity measurements indicated a rather bicontinuous structure, the TS model is the most realistic model that can be used in this case. The scattering curves in dependence of [bmim][BF$_4$] content are shown at 30°C and 60°C in Figure IV-60 a and b and at 90°C and 150°C in Figure IV-61 a and b, respectively. For all curves the scattering intensity decreases with increasing RTIL content, because of the unfavorable contribution of the BF$_4^-$ anion to the scattering contrast. In general, the scattering intensity was very low.

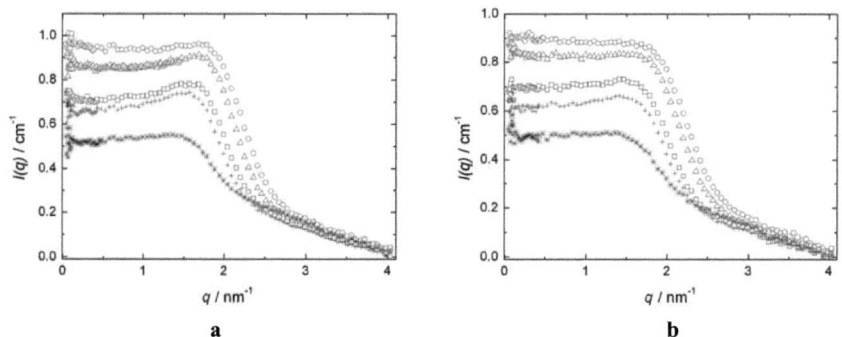

Figure IV-60. SANS curves at 30°C (a) and 60°C (b) along the experimental path with P_S= 0.65 for different [bmim][BF$_4$] weight fractions (0% (○), 3 % (Δ), 9 % (□), 12 % (+), 15 % (∗)).

At 30°C, for low RTIL concentration a slight scattering peak could be observed. With increasing temperature, the peak becomes even less pronounced and disappears completely at 90°C and 150°C, respectively. Therefore, the TS model did not yield convenient fits.

IV.2.6. [bmim][BF$_4$] in high temperature stable microemulsions

Furthermore, no clear Porod limit could be observed within the measured temperature and concentration range. Hence, a convenient data analysis with the models described in literature was not possible. Nevertheless, the existence of colloidal structures between 30°C and 150°C was confirmed.

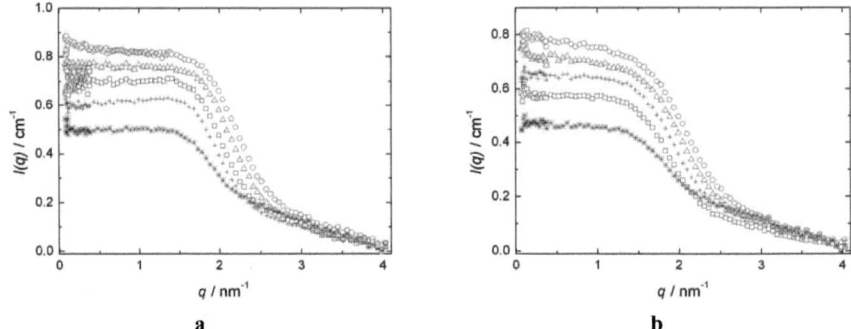

Figure IV-61. SANS curves at 90°C (a) and 150°C (b) along the experimental path with $P_S = 0.65$ for different [bmim][BF$_4$] weight fractions (0% (○), 3 % (Δ), 9 % (□), 12 % (+), 15 % (∗)).

2.6.3. Concluding remarks

In conclusion, it has been demonstrated that [bmim][BF$_4$] is suitable to formulate high-temperature stable nonaqueous microemulsions. Compared to the protic ionic liquid EAN, a significantly higher amount of surfactant is necessary to obtain a single phase microemulsion. Consequently, the single phase area is much smaller compared to microemulsions with EAN. Due to the high amount of surfactant a microemulsion structure with well separated RTIL droplets stabilized by surfactant and cosurfactant in a continuous oil matrix could not be obtained for [bmim][BF$_4$]. Conductivity measurements prove the high thermal stability of the microemulsions, a percolation phenomenon could not be observed. The specific conductivity increases continuously with increasing amount of RTIL and with increasing temperature for a constant [bmim][BF$_4$] weight fraction. Compared to neat [bmim][BF$_4$] the specific conductivity is much lower for any point in the phase diagram and temperature. Furthermore, the specific conductivity is remarkably higher compared to pure dodecane and to microemulsions with EAN. For these reasons, a rather bicontinuous structure could be supposed. Temperature dependent SANS experiments confirmed the high thermal stability of these microemulsions. The correlation peak at 30°C was not very pronounced and decreased with increasing amount of RTIL. With increasing temperature, the scattering peak became weaker and disappeared completely above 90°C. Hence, the TS model could not be used to fit the curves in a convenient way. Nevertheless, a remarkable change in structure depending on the nature of the polar RTIL was observed. Protic ionic liquids seem to be suitable to formulate droplet like microemulsions, while the imidazolium based [bmim][BF_4] yields less defined structures.

3. Alkali oligoether carboxylates – a new class of ionic liquids

3.1. Abstract

In the present chapter a new class of Ionic Liquids (ILs), the "alkali oligoether carboxylates" is introduced. The combination of simple alkali ions and an oligoether carboxylate leads to the formation of ILs of the general type R-O-$(CH_2CH_2O)_3$-CH_2-COOX, where R is a short alkyl chain and X an alkali cation. The synthesis strategy and physicochemical properties of the salts are presented. The described substances are promising materials due to their pronounced electrochemical and thermal stability. Further, it was found that the cytotoxicity of such "simple" alkali carboxylate ionic liquids is very low. In addition, possible reasons for the formation of ionic liquids with alkali metals are proposed.

3.2. Introduction

To date, little attention has been paid to systems of ionic liquids involving small inorganic cations. Rees *et al.* investigated monomeric barium bisalkoxides with the general formula Ba[O$(CH_2CH_2O)_n CH_3]_2$ that were reported to be liquid at ambient temperature.[267] Thereby, a solution structure with a coordinative saturation at the central metal atom with oxygen atoms was proposed, implying an intramolecular complexation of the metal ions similar to the coordination prevailing in macrocyclic polyethers.[268]

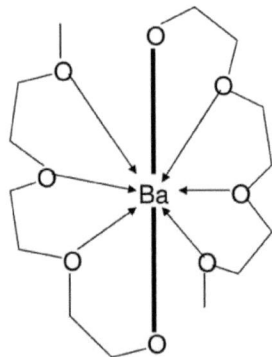

Figure IV-62. Calm shell oligoether bisalkaloxide, redrawn from ref. 267.

IV. Results and discussion

The calm shell structure for the barium salt of the oligoether bisalkaloxide proposed by Rees et al.[267] is illustrated in Figure IV-62.

Other studies describe the synthesis and characterization of polyether carboxylates of heavy alkaline-earth metals,[269] which were found to be viscous oils that transform into hygroscopic solids upon reduction of the water content. Likewise, the yttrium(III) salt of 3,6,9-trioxadecanoic acid (TOD) was characterized as an extremely hygroscopic viscous oil that could only be isolated as trihydrate.[270] Recently, Justus et al. reported novel ionic liquids consisting of trialkylammoniododecaborate anions and counter-ions including lithium and potassium.[271] Imidazolium cation-based ionic liquids with poly(ethyleneglycol) moieties[272] and fluorinated anions or ether and alcohol functional groups[273] with halide and fluorinated anions have also been subject to studies. Pernak et al. described dialkoxymethyl-substituted imidazolium cations combined with [BF$_4$] and [NTf$_2$] anions,[274] while ether-derivatized imidazolium halides were reported by Fei et al.[275] In the present study, ionic liquids based on small inorganic cations and oligoether carboxylate anions were successfully synthesized. We present for the first time a new family of ILs comprising alkali cations and 2,5,8,11-tetraoxatridecan-13-oate (TOTO) as anion, as shown in Figure IV-63.

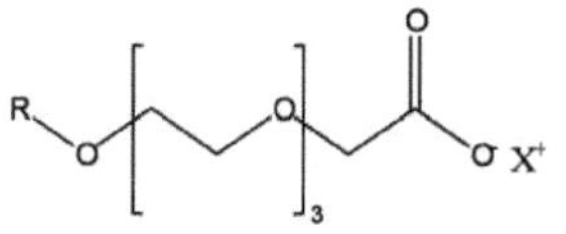

Figure IV-63. Structure of alkali 2,5,8,11- tetraoxotridecane-13-oate with X = Li$^+$, Na$^+$, K$^+$.

We investigate here the physicochemical properties of the salts, and propose possible reasons for the formation of ionic liquids with alkali metals.

3.3. Results and discussion

3.3.1. Synthesis

2,5,8,11-Tetraoxatridecan-13-oic acid (TOTOA) was prepared according to a modified procedure described by Matsushima et al.,[197] in high purity (> 99%, GC). Ionic liquids were easily obtained by neutralizing the acid with an equimolar amount of alkali hydroxide or hydrogencarbonate in aqueous solution. Water was first removed by lyophilization, followed

IV.3. Alkali oligoether catboxylates – a new class of ionic liquids

by drying in vacuum. Both the lithium and the sodium salt were obtained as colorless or faint yellow liquids at room temperature, while the potassium carboxylate was a white solid. Details of the synthesis are given in the experimental section (III.2.2.). The following scheme summarizes the synthesis strategy.

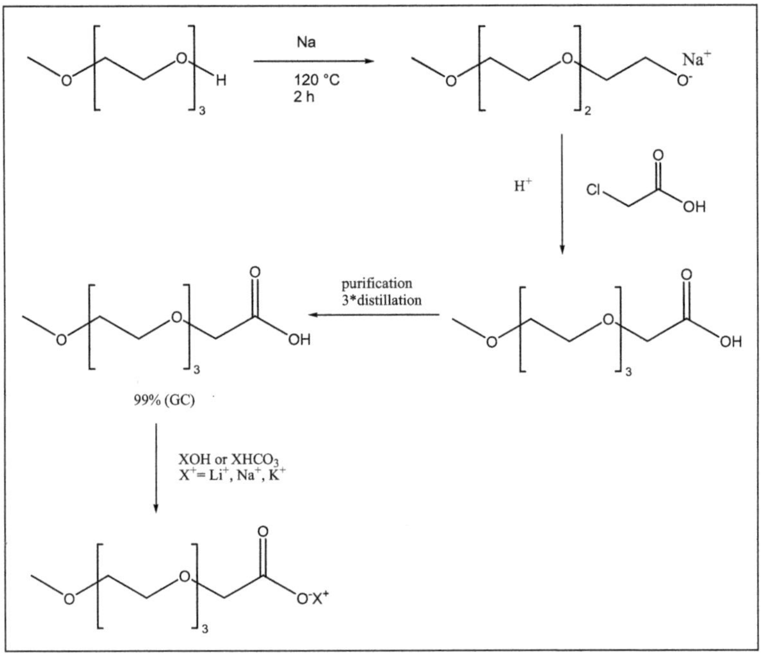

Figure IV-64. Synthesis strategy of the TOTO alkali salts.

The water content of the hygroscopic salts was determined by means of coulometric Karl-Fischer titration and found to be below 300 ppm (m/m) for the room-temperature liquid salts. All ionic liquids were characterized by ^{1}H and ^{13}C-NMR spectroscopy, mass spectrometry as well as by elemental analysis. Melting and glass transition temperatures were measured by differential scanning calorimetry (DSC), decomposition temperatures by thermogravimetric analysis (TGA) (more details are given in the experimental section). [Li]- and [Na][TOTO] show glass transitions, the temperature of which was determined from the intersection of the curve and the half-way line between the two baselines. The water content, melting points, θ_m, glass transition temperatures, θ_g, and decomposition temperatures, θ_d, are summarized in

Table IV-14.

Table IV-14. Physical properties of the alkali salts of 2,5,8,11-tetraoxatri-decan-13-oic acid.

Cation	H$_2$O / ppm	θ_m^a resp. θ_g^b / °C	θ_d / °C
Li$^+$	103	-53^b	357
Na$^+$	211	-57^b	384
K$^+$	1292	+60^a	369

θ_m^a: melting point, θ_g^b: glass transition temperature, θ_d: decomposition temperature

Generally, the thermal stability of ILs is very sensitive to the type of both the cation and the anion. For instance, imidazolium cations tend to be thermally more stable than tetraalkylammonium species. Regarding the anions, a series of relative stability can be established ranking from [PF$_6^-$] over [BF$_4^-$] to halides.[276] All TOTOA alkali salts exhibit excellent thermal stability. The decomposition temperatures of the three substances are very similar, indicating that the nature of the cation plays a minor role in this context.

3.3.2. Conductivity and viscosity

As [K][TOTO] was solid at ambient temperature and the viscosity of [Li][TOTO] was too high, it was not possible to obtain convenient data for the viscosity and conductivity of the salts. Therefore, in the following only viscosities and conductivities of [Na][TOTO] are presented. Conductivity measurements were carried out under nitrogen atmosphere at temperatures ranging from (25 to 145)°C. Viscosities were measured under argon atmosphere at defined temperatures between (25 and 65)°C, revealing Newtonian behavior over the whole range. The Newtonian behavior is shown exemplarily for 45°C, data of the shear stress, τ, versus the shear rate, $\dot{\gamma}$ exhibit a linear relation, the full line represents a linear fit.

IV.3. Alkali oligoether catboxylates – a new class of ionic liquids

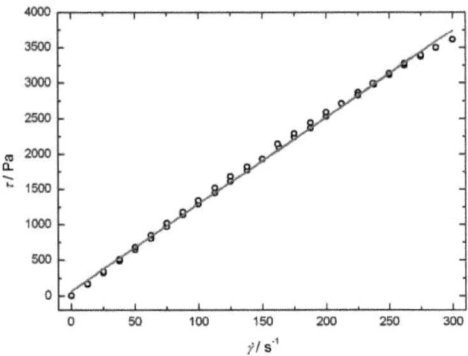

Figure IV-65. Shear stress, τ, versus shear rate, $\dot{\gamma}$, demonstrating the Newtonian behavior exemplarily shown for [Na][TOTO] at 45°C, the straight line represents a linear fit.

For the sodium salt, a characteristic increase in specific conductivity and decrease in viscosity with increasing temperature was detected. Both the conductivity and viscosity data were found to be well described by the empirical VFT equation (VFT) of the following form

$$\kappa = \kappa_0 \exp(-B/(T-T_0)) \qquad \text{(IV-21)}$$

$$\eta = \eta_0 \exp(B/(T-T_0)) \qquad \text{(IV-22)}$$

with the fit parameters A, the so-called VFT temperature T_0 and κ_0 and η_0, respectively. The corresponding plot and the fit results are given in Figure IV-66 and Table IV-15, respectively.

To calculate the molar conductivity, Λ, of [Na][TOTO] densities, ρ, were measured between (25 and 65)°C with a vibrating-tube densimeter. The obtained values for ρ and Λ are summarized in Table IV-16.

IV. Results and discussion

Table IV-15. VFT parameters of conductivity and viscosity data for [Na][TOTO].

	κ_0 / S m^{-1}	B / K	T_0 / K
[Na][TOTO]	33.7	1485	193
	η_0 / P	B / K	T_0 / K
[Na][TOTO]	0.0024	1140	213

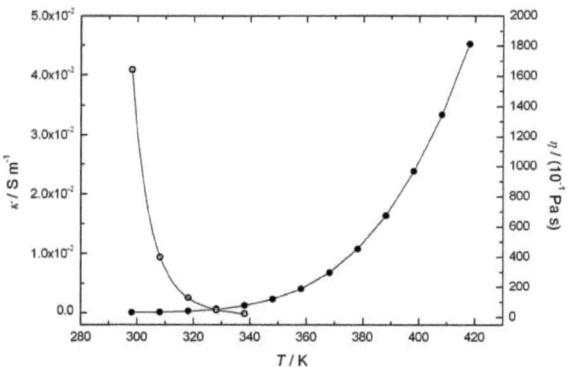

Figure IV-66. Specific conductivities, κ, (●) and viscosities, η, (○) of the sodium salt as a function of temperature, full lines represent fits with the VFT equation.

Table IV-16. Densities, ρ, and molar conductivities, Λ, of [Na][TOTO].

θ / °C	ρ / kg m^{-3}	$10^4 \Lambda$ / S m^2 mol^{-1}
25	1247.44	9.08 10^{-5}
35	1242.68	2.28 10^{-4}
45	1237.92	5.55 10^{-4}
55	1233.15	1.21 10^{-3}
65	1228.39	2.44 10^{-3}

The relation of fluidity to conductance can be considered in terms of a Walden plot of the data, as described by Angell and coworkers as discussed in detail in section II.1.2.2[54,55,56] The

IV.3. Alkali oligoether catboxylates – a new class of ionic liquids

Walden plot obtained for the sodium salt over a temperature range of (25 – 65)°C is shown in Figure IV-67.

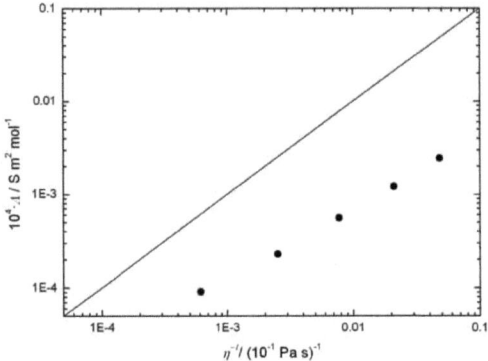

Figure IV-67. Walden plot for [Na][TOTO] (●) at temperatures ranging from (25 - 65)°C, the solid line represents the ideal KCl line.

It is obvious that the curve found for [Na][TOTO] lies significantly below the ideal KCl line. MacFarlane and coworkers[57] proposed to term systems exhibiting such behavior "Liquid Ion Pairs". The divergence of the single values obtained for [Na][TOTO] at different temperatures from the ideal line is summarized in Table IV-17, where ΔW is the vertical deviation ($\Delta W = \log \Lambda - \log \eta^{-1}$).

Table IV-17. Deviation ΔW of [NA][TOTO] from the ideal KCl line in the Walden plot at different temperatures.

θ /°C	25	35	45	55	65
ΔW	0.8	1.0	1.1	1.2	1.3

These findings clearly suggest the presence of strong ion pairing in the liquid sodium oligoether carboxylate.

3.3.3. Electrochemical stability

Electrochemical stability was studied using cyclic voltammetry (CV) with Pt working and counter electrodes vs. an Ag/Ag$^+$/Kryptofix reference, according to Izutsu.[276] The CV measurements have been performed by Dr. Christian Schreiner who is gratefully acknowledged at this place.

Generally speaking, the electrochemical window of an ionic liquid is defined by the reduction of the cation and oxidation of the anion. The width of this window is quite high for many ILs, often exceeding 4 V.[277] The cyclic voltammogram of [Na][TOTO], was recorded first in anodic direction with a scan rate of 10 mV s^{-1}. The cathodic and anodic limits are about -2.0 V and 2.7 V vs. Ag/Ag$^+$, respectively, resulting in an electrochemical window of 4.7 V. The CV data were recorded using a 0.55 M solution of [Na][TOTO] in acetonitrile without added inert salt, giving rise to the appearance of traces of impurities such as the small peak seen in the anodic branch. All experiments were carried out in an inert gas atmosphere. For [Li][TOTO], an outstandingly wide electrochemical window of 6.7 V was found, with a cathodic limit of about -3.3 V and an anodic limit of about 3.4 V. The cyclovoltammograms of [Na][TOTO] and [Li][TOTO] are shown in Figure IV-68 a and b, respectively. These findings indicate an extraordinary high electrochemical stability of the as-synthesized alkali oligoether carboxylates.

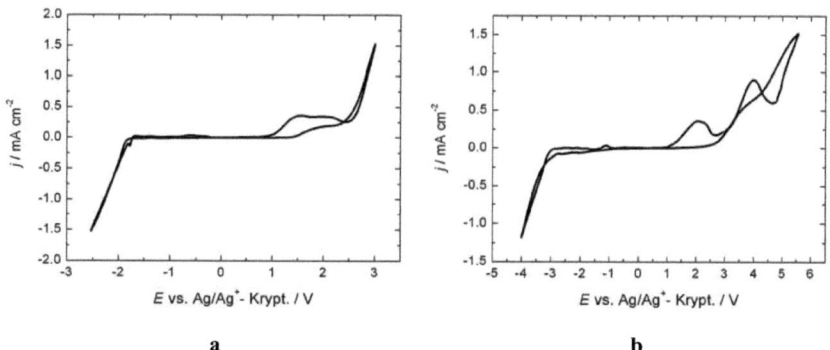

Figure IV-68. Cyclovoltammogram of [Na][TOTO] a) and [Li][TOTO] b), respectively.

For comparison, the electrochemical windows for some typical ionic liquids are illustrated in Figure IV-69.

IV.3. Alkali oligoether catboxylates – a new class of ionic liquids

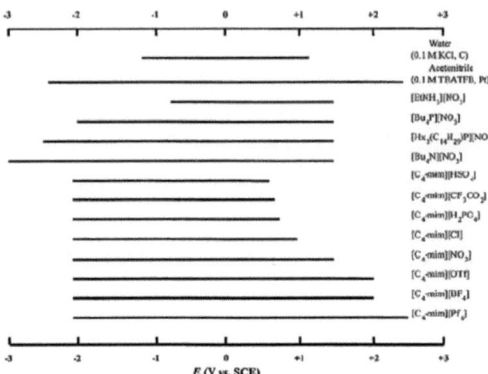

Figure IV-69. Electrochemical window for some typical ionic liquids, reproduced from ref. 2.

3.3.4. Cytotoxity tests

To look further to potential applications, the cytotoxicity of the new ionic liquids was studied using a MTT assay with Hela cells. Hela cells received from the American Type Culture Collection (ATCC). MTT (3-(4,5-dimethylthiazol-2-yl)-2,5-diphenyl tetrazolium bromide) assays were conducted following a procedure proposed by Mosmann.[278] (for details see experimental section). I am grateful to Eva Maurer who performed the cytotoxicity tests. As a reference, some common ILs were tested likewise, namely ethylammonium nitrate (EAN) as well as the two imidazolium-based cations [emim] (1-ethyl-3-methylimidazolium) and [bmim] (1-butyl-3-methyl-imidazolium) with the anions tetrafluoroborate and ethylsulfate. For each sample, the IC_{50} value was determined, which represents the concentration of test substance that reduces cell viability by 50% compared to the untreated control. Thus, the higher the IC_{50} value, the less toxic is the substance. All experiments were repeated four times; the resulting averaged IC_{50} values are plotted in Figure IV-70. From these data, two important conclusions can be drawn. First, the cytotoxicity of the alkali TOTO salts depends on the nature of the alkali cation, with [Na][TOTO] exhibiting the highest IC_{50} value and therefore being least toxic. Second, the IC_{50} values of the TOTO salts are significantly higher than those of the widely studied imidazolium-based ILs. In this context, [Na][TOTO] is for example fifty times less cytotoxic than [bmim][BF$_4$]. In fact, the cytotoxicity of the alkali TOTO salts was found to be comparable to that of EAN. This finding underlines the potential of these new ionic liquids in view of manifold applications.

IV. Results and discussion

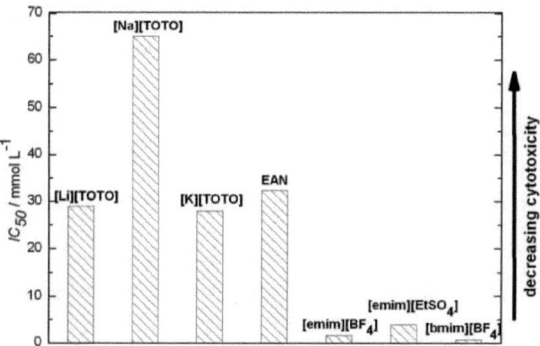

Figure IV-70. Cytotoxicity of the TOTO alkali salts as compared to common ILs.

3.4. Concluding remarks

In conclusion, a new family of ionic liquids was introduced based on the combination of simple alkali ions and oligoether carboxylates. The described substances are promising materials due to their pronounced electrochemical and thermal stability. The concept of the ionicity plot was successfully applied to the sodium salt for which strong ion pairing was observed. It is evident that these new ionic liquids are featured by interesting physicochemical properties, including remarkably high thermal and electrochemical stability. Noteworthy, the lithium and sodium TOTO salts are liquid at room temperature. So far room-temperature ILs based on alkali cations have not been described in literature. Nonetheless, the reason for these substances being liquid at ambient temperature remains to be discussed. The interactions of alkali and alkaline-earth ions with the CH_2CH_2O unit in solution have been studied extensively. The complexation of alkali ions by cyclic and acyclic polyether ligands has been reviewed.[279] It is furthermore well known that the so-called "glymes" (polyethylenglycol dimethyl ethers) with the general structure $CH_3O(CH_2CH_2O)_nCH_3$ possess a high affinity to alkali cations. Interestingly, in particular the tetraglyme (TeG, n = 4) is structurally very similar to the TOTO anion. Grobelny *et al.* conducted a comparative study on solutions of potassium ions in tetraglyme in the presence and absence of 18-crown-6.[280] It was claimed that tetraglyme is a strong complexing agent for K^+, the latter being surrounded by the tetraglyme molecules in a pseudo-crown ether fashion. Other studies reported significant interactions of glymes with Li^+ in solution.[281] Ab initio calculations for the tetra-, penta- and hexaglyme complexes of lithium led to the conclusion that the coordination number for Li^+

- 155 -

IV.3. Alkali oligoether catboxylates – a new class of ionic liquids

with respect to the solvent oxygen atoms varies between 4 and 6.[282]

Based on this, we suggest that in the present case a complexation of the alkali cations by their combined oligoether-carboxylate counter-ion is effective to result in room-temperature liquid salts. Simultaneous "intramolecular" charge neutralization and complexation apparently diminishes the salt character of the substance, thereby shifting it to the state of a neutral "molecule" and correspondingly lowering the melting point. Such a hypothesis is supported by the Walden plot giving evidence for the formation of liquid ion pairs.

Further, it was shown that the cytotoxicity of such "simple" alkali carboxylate ionic liquids is very low. The physicochemical properties and the design of further new room-temperature liquid TOTOA derivatives are currently under investigation. In fact, the use of simple tri- and tetraalkylammonium instead of alkali ions have also yielded ionic liquids at ambient temperature which appear to be featured by additional desirable properties, including low viscosity. Combination of the TOTO anion with "biological" cations such as choline might pave the way to still "greener" ILs.

V. Summary

The thesis can be subdivided into three main parts:

i. Conductivity studies of the anion effect on imidazolium based ionic liquids over a wide temperature range (-25-195)°C

ii. Formulation of high temperature stable nonaqueous microemulsions with room temperature ionic liquids as polar phase

iii. Synthesis and characterization of new ionic liquids based on alkali cations

i. In the first part conductivities, κ, of four different highly pure imidazolium based room temperature ionic liquids (RTILs) have been studied within a temperature range between (-25 to 195)°C. Thereby, the cationic scaffold, the 1-butyl-3-methylimidazolium cation ([bmim$^+$]), was kept constant while the anions were varied. The investigated anions were dicyanamide ([DCA$^-$]), hexafluorophosphate ([PF$_6^-$]), trifluoroacetate ([TA$^-$]) and trifluoromethanesulfonate ([TfO$^-$]). It is quite surprising that studies of important physicochemical transport properties are still scarce in the field of ionic liquids. At a given temperature the conductivity decreased in the order [bmim][DCA] > [bmim][TA] > [bmim][TfO] > [bmim][PF$_6$]. Temperature dependence of the conductivity could be well described by the empirical Vogel-Fulcher-Tammann equation. Whilst our data compare favorably with some literature results, significant deviations from others were noted. To calculate the molar conductivity, Λ, of the RTILs densities, ρ, were measured between (5 and 65)°C. Walden plots of the molar conductance, available for [bmim][PF$_6$], [bmim][TfO] and [bmim][TA] in the limited temperature range of (5 to 65)°C, suggest that these RTILs can be classified as high-ionicity ionic liquids. This study on the anion dependence of RTIL conductivity complements a companion paper on the cation dependence, particularly the alkyl chain length of 1-alkyl-3-methylimidazolium, of κ for a series of imidazolium tetrafluoroborate salts.

The results are summarized in a manuscript entitled "The Conductivity of Imidazolium-Based Ionic Liquids from (248 to 468) K. B. Variation of the Anion.", submitted for publication in J.

V. Summary

Chem. Eng. Data. (2009) (Authors: Oliver Zech, Alexander Stoppa, Richard Buchner, and Werner Kunz)

The companion study performed by Alexander Stoppa is part of a manuscript entitled "The Conductivity of Imidazolium-Based Ionic Liquids from (-35 to 195)°C. A: Variation of Cation`s N-Alkyl Chain.", accepted for publication in J. Chem. Eng. Data.(2009) (Authors: Alexander Stoppa, Oliver Zech, Werner Kunz, Richard Buchner)

ii. The mainpart of the thesis is focused on the application of RTILs in nonaqueous, high temperature stable microemulsions. For this purpose two different RTILs have been chosen, namely ethylammonium nitrate (EAN) and 1-butyl-3-methylimidazolium tetrafluoroborate ([bmim][BF$_4$]). EAN is probably the most popular example for a protic ionic liquid, it is highly polar and shows many similarities to water. On the contrary, [bmim][BF$_4$] is a typical representative for an aprotic RTIL. Compared to EAN, the polarity is reduced and similarities to water are scarce. Furthermore, the cationic long chain ionic liquid surfactant 1-hexadecy-3-methylimidazolium chloride ([C$_{16}$mim][Cl]) was used, while decanol acted as cosurfactant and dodecane as apolar phase. Interestingly, the two RTILs showed a completely different topology in the pseudo-ternary phase diagrams. While the phase boundaries for microemulsions with [bmim][BF$_4$] were regular, irregular phase boundaries could be reported for microemulsions with EAN. The area of the clear and isotropic single phase region was significantly reduced in the case of [bmim]BF$_4$]. Consequently, the amount of surfactant necessary to form a true Schulman type microemulsion in the case of [bmim][BF$_4$] is remarkably higher. Before studying the thermal stability of these microemulsions, a comprehensive characterization at room temperature was necessary. Therefore, conductivity, dynamic light scattering (DLS), viscosity and small angle X-ray (SAXS) measurements have been performed.

For the microemulsions with EAN as polar phase, a percolative behavior in conductivity has been observed, a model of dynamic percolation could be applied. DLS experiments confirmed the existence of spherical EAN droplets stabilized by surfactant with dodecane as continuous phase. A swelling of the reverse micelles with increasing amount of EAN could be noted. The swelling behavior was further confirmed by SAXS measurements, since the domain size extracted from the Teubner-Strey (TS) model increased with increasing amount of EAN. Further, from the Porod limit and the experimental invariant, data of microemulsions with

EAN were in agreement with models for spherical ionic liquid in oil (IL/o) aggregates.

On the contrary, for [bmim][BF$_4$], no percolation phenomenon could be observed neither with conductivity nor with viscosity measurements, since these transport properties increased continuously with the [bmim][BF$_4$] weight fraction. The autocorrelation functions extracted from DLS measurements did not allow a convenient data analysis. SAXS data confirmed the existence of well-structured microemulsions. From the characterization methods used, a rather bicontinuous structure in the case of [bmim][BF$_4$] was assumed. These investigations at ambient temperature demonstrate that the nature of the polar RTIL plays an important role in microemulsion phase behavior and structure.

The results obtained from the characterization at ambient temperature form the basis of a publication entitled "Microemulsions with an Ionic Liquid Surfactant and Room Temperature Ionic Liquids as Polar Pseudo-phase" published in J. Phys. Chem. B (2009) (Authors: Oliver Zech, Stefan Thomaier, Pierre Bauduin, Thomas Rück, Didier Touraud, and Werner Kunz).

Since microemulsions with EAN as polar phase yielded more promising results concerning the microemulsion structure for potential applications, their thermal stability has been investigated. Visually, no phase change within 30°C and 150°C could be observed. Temperature dependent conductivity measurements demonstrated the high thermal stability in general and gave information about the microstructure in particular. A percolative behavior was detected over the whole investigated temperature range (30 - 150)°C. Interestingly, temperature affected the percolation significantly. The higher the temperature, the less of RTIL can be integrated into the reverse micelles before percolation takes place. DLS measurements at ambient temperature indicated a swelling of the reverse micelles with increasing amount of RTIL below the percolation threshold. Above this threshold, the formation of elongated structures or droplet clusters could be assumed. Temperature dependent SANS experiments further confirmed the existence of spherical structures with RTIL cores even at 150°C. A swelling of these droplets could be observed with increasing temperature and with increasing amount of EAN. To the best of my knowledge, this is the first study dealing with high temperature stable microemulsions at ambient pressure.

The concept of the formulation of high temperature stable microemulsions with RTIL in

V. Summary

conjunction with conductivity measurements and preliminary SANS experiments are part of a manuscript entitled "Ionic Liquids in Microemulsions – a Concept to Extend the Conventional Thermal Stability Range of Microemulsions" accepted for publication in Chem. Eur. J. (2009) (Authors: Oliver Zech, Stefan Thomaier, Agnes Kolodziejski, Didier Touraud, Isabelle Grillo, and Werner Kunz).

Details on temperature dependent SANS experiments for various points in the phase diagram in combination with DLS measurements form the basis of a manuscript entitled "Ethylammonmium nitrate in High Temperature Stable Microemulsions" submitted to J. Phys. Chem. B (2009) (Authors: Oliver Zech, Stefan Thomaier, Agnes Kolodziejski, Didier Touraud, Isabelle Grillo, and Werner Kunz).

Moreover, the effect of the surfactant alkyl chain length ($[C_n mim][Cl]$, with n = 14, 16, 18) on structure, thermal stability and phase behavior of microemulsions with EAN as polar phase has been investigated. It could be demonstrated that the chain length of the surfactant plays an important role on phase diagram topology, phase behavior and flexibility of the interfacial film. In order to obtain comparable systems, all conditions, except the surfactant chain length, were kept identical. On the one hand, the area of the one phase region in the pseudo-ternary phase diagram increased from C_{14} to C_{18}. On the other hand, the rigidity of the interfacial film raised with increasing surfactant chain length. For all systems, irregular phase boundaries have been found. Conductivity measurements demonstrated a percolative behavior for all three microemulsion systems. For the C_{14} and C_{16} chain length, the temperature dependent percolation threshold volume fractions, ϕ_P, were very similar at a given temperature. For the C_{18} chain length ϕ_P was significantly higher at 30°C, 60°C and 90°C. These findings underline the increasing rigidity of the interfacial film for microemulsions with the C_{18} chain length. Viscosity measurements did not show any percolation effect for all three systems, the viscosities were all in the same order of magnitude for a given EAN content. DLS experiments confirmed a swelling behavior with increasing amount of EAN for samples with EAN weight fractions below the percolation threshold. Above this threshold, a bimodal decay in the intensity autocorrelation function was observed indicating the formation of non-spherical structures or droplet clusters. No significant effect in size with increasing chain length from the DLS measurements at a given EAN content could be reported.

V. Summary

Additionally, it was demonstrated that [bmim][BF$_4$] is suitable to formulate high-temperature stable nonaqueous microemulsions. Compared to the protic ionic liquid EAN, a significantly higher amount of surfactant is necessary to obtain a single phase microemulsion. Due to the high amount of surfactant a microemulsion structure with well separated RTIL droplets stabilized by surfactant and cosurfactant in a continuous oil matrix could not be obtained. On the one hand, conductivity measurements prove the high thermal stability of the microemulsions, on the other hand they suggest less defined structures since no percolation phenomenon could be detected. Temperature dependent SANS experiments confirmed the high thermal stability of these microemulsions. The correlation peak at 30°C was not very pronounced and decreased with increasing amount of RTIL. With increasing temperature the scattering peak was attenuated and disappeared completely above 90°C. Hence, the TS model could not be used to fit the curves in a convenient way. The scattering intensity did not follow a q^{-4} decay at large q values. Consequently, no Porod limit could be extracted. Furthermore, due to the high amount of surfactant the GIFT method was not applicable. Nevertheless, SANS experiments confirmed the existence of colloidal structures over a wide temperature range. However, a remarkable change in structure depending on the nature of the polar RTIL was observed. Protic ionic liquids seem to be suitable to formulate droplet like microemulsions, while the imidazolium based [bmim][BF$_4$] yielded less defined structures.

Finally, the oil phase dodecane has been replaced by sustainable oil, namely biodiesel. The phase diagram observed with biodiesel exhibits a huge clear and isotropic single phase region comparable to microemulsions with dodecane. The viscosity of the microemulsions was in the same order of magnitude compared to dodecane as continuous phase. Conductivity measurements indicated percolation phenomena when the amount of EAN was increased. Contrary to similar microemulsions with dodecane, the percolation threshold seemed to be relatively insensitive to temperature. Consequently, ϕ_P did not change significantly with increasing temperature. SAXS experiments confirmed the existence of IL/o spheres. These microemulsions made of sustainable oil provide a way in the formulation of greener microemulsions.

This study forms the basis of a manuscript entitled "Biodiesel, a Sustainable Oil, in High Temperature Stable Microemulsions Containing a Room Temperature Ionic Liquid as Polar Phase" submitted to Energy & Environmental Science (2009) (Authors: Oliver Zech, Pierre Bauduin, Peter Palatzky, Didier Touraud, and Werner Kunz).

V. Summary

iii. In the third part, a new class of ionic liquids (ILs), the "alkali oligoether carboxylates" was introduced. The combination of simple alkali ions and oligoether carboxylates leads to the formation of ILs of the general type R-O-$(CH_2CH_2O)_3$-CH_2-COOX, where R is a short alkyl chain and X an alkali cation. These substances are promising materials due to their pronounced electrochemical and thermal stability. The concept of the "ionicity plot" was successfully applied to the sodium salt for which strong ion pairing was observed. Based on this, it was suggested that a complexation of the alkali cations by their combined oligoether-carboxylate counter-ion is effective to result in room-temperature liquid salts. Furthermore, the cytotoxicity of these substances is relatively low compared to common imidazolium based ILs. So far, room-temperature ILs based on alkali cations have not been described in literature.

The results form the basis of a paper entitled "Alkali Oligoether Carboxylates - a New Class of Ionic Liquids" published in Chem. Eur. J. (2009) (Authors: Oliver Zech, Matthias Kellermeier, Stefan Thomaier, Eva Maurer, Regina Klein, and Werner Kunz).

Furthermore this work is part of a patent entitled "Onium Salts of Carboxyalkyl-Terminated Polyoxyalkylenes for Use as High-Polar Solvents and Electrolytes" PCT Int. Appl. (2008) (Authors: Werner Kunz, Stefan Thomaier, Eva Maurer, Oliver Zech, Matthias Kellermeier, Regina Klein).

VI. Appendix

1. List of publications

1. Oliver Zech, Stefan Thomaier, Pierre Bauduin, Thomas Rück, Didier Touraud, and Werner Kunz, "Microemulsions with an ionic liquid surfactant and room temperature ionic liquids as polar pseudo-phase", J. Phys. Chem. B, *113(2)* (2009) 465-473.

2. Oliver Zech, Matthias Kellermeier, Stefan Thomaier, Eva Maurer, Regina Klein, Christian Schreiner, and Werner Kunz, "Alkali oligoether carboxylates - a new class of ionic liquids", Chemistry - A European Journal *15 (6)* (2009) 1341-1345.

3. Oliver Zech, Stefan Thomaier, Agnes Kolodziejski, Didier Touraud, Isabelle Grillo, and Werner Kunz, "Ionic Liquids in microemulsions – a concept to extend the conventional thermal stability range of microemulsions", Chemistry - A European Journal, *16 (3)* (2010) 783-786.

4. Sekh Mahiuddin, Oliver Zech, Sabine Raith, Didier Touraud, and Werner Kunz, "Catanionic micelles as model to mimic biological membranes in the presence of anastetic alcohols", Langmuir, *25 (21)* (2009) 12516-12521.

5. Alexander Stoppa, Oliver Zech, Werner Kunz, and Richard Buchner, "Electrical conductivities of imidazolium based ionic liquids over a wide temperature range, part A: The effect of chain length", J. Chem. Eng. Data, in press.

6. Oliver Zech, Alexander Stoppa, Richard Buchner, and Werner Kunz, "Electrical conductivities of imidazolium based ionic liquids over a wide temperature range, part B: The effect of counter ion", J. Chem. Eng. Data, in press.

7. Oliver Zech, Stefan Thomaier, Agnes Kolodziejski, Didier Touraud, Isabelle Grillo, and Werner Kunz, "Ethylammonium nitrate in high temperature stable microemulsions", J. Colloid Interface Sci., in press

8. Oliver Zech, Pierre Bauduin, Peter Palatzky, Didier Touraud, and Werner Kunz, "Biodiesel, a sustainable oil, in high temperature stable microemulsions containing room temperature ionic liquid as polar phase", Energy Environmental Sci., accepted manuscript

VI. Appendix

2. Patent

Werner Kunz, Stefan Thomaier, Eva Maurer, Oliver Zech, Matthias Kellermeier, Regina Klein , "Onium salts of carboxyalkyl-terminated polyoxyalkylenes for use as high-polar solvents and electrolytes" PCT Int. Appl. (2008), 19pp. CODEN: PIXXD2 WO 2008135482 A2 20081113.

VII. Literature Cited

[1] P. Walden, Molecular weights and electrical conductivity of several fused salts, *Bull. Acad. Sci. St. Petersburg* **1914**, 405-422.

[2] N. V. Plechkova, K. R. Seddon, Applications of ionic liquids in the chemical industry, *Chem. Soc. Rev.* **2008**, *37*, 123-150.

[3] R. D. Rogers, K. R. Seddon, Ionic liquids: Industrial applications to green chemistry, ed. R. D. Rogers, K. R. Seddon, ACS Symp. Ser., American Chemical Society, Washington D.C. 2002, vol. 818.

[4] M. J. Earle, K. R. Seddon, Ionic liquids. Green solvents for the future, *Pure Appl. Chem.* **2000**, *72*, 1391-1398.

[5] M. Freemantle, Designer solvents-ionic liquids may boost clean technology development, *Chem. Eng. News* **1998**, *76*, 32-37.

[6] A. J. Carmichael, M. Deetlefs. M. J. Earle, U. Fröhlich and K. R. Seddon, Ionic liquids as green solvents: Progress and prospects, ed. R. D. Rogers, K. R. Seddon, ACS Symp. Ser., American Chemical Society, Washington D.C. 2003, vol. 856, 14-31.

[7] F. van Rantwiijk, R.A. Sheldon, Biocatalysis in ionic liquids, *Chem. Rev.* **2007**, *107*, 2757-2785.

[8] V. I. Pârvulescu, C. Hardacre, Catalysis in ionic liquids, *Chem. Rev.* **2007**, *107*, 2615-2665.

[9] M. Haumann, A. Riisager, Hydroformylation in room temperature ionic liquids (RTILs): Catalyst and process developments, *Chem. Rev.* **2008**, *108,* 1474-1497.

[10] M. A. P. Martins, C. P. Frizzo, D. N. Moreira, N. Zanatta, H.G. Bonacorso, Ionic liquids in heterocyclic synthesis, *Chem. Rev.* **2008**, *108*, 2015-2050.

[11] L. A. Blanchard, D. Hancu, E. J. Beckman, J. F. Brenecke, Green processing using ionic liquids and CO_2, *Nature* **1999**, *399*, 28-29.

VII. Literature cited

[12] N. Byrne, P. C. Howlett, D. R. MacFarlane, M. Forsyth, The zwitterion effect in ionic liquids: towards practical rechargeable lithium-metal batteries, *Adv. Mater.* **2005**, *17*, 2497-2501.

[13] P. Wang, S. M. Zakeeruddin, J.-E. Moser, M. Graetzel, A new ionic liquid electrolyte enhances the conversion efficiency of dye-sensitized solar cells, *J. Phys. Chem. B* **2003**, *107*, 13280-13285.

[14] N. Yamanaka, R. Kawano, W. Kubo, T. Kitamura, Y. Wada, M. Watanabe, S. Yanagida, Ionic liquid crystal as a hole transport layer of dye-sensitized solar cells, *Chem. Commun.* **2005** 740-742.

[15] P. Hapiot, C. Lagrost, Electrochemical reactivity in room-temperature ionic liquids, *Chem. Rev.* **2008**, *108*, 2238-2264.

[16] P. Wasserscheid, W. Keim, Ionic liquids – new "solutions" for transition metal catalysis, *Angew. Chem. Int. Ed.* **2000**, *39*, 3772-3789.

[17] T. Welton, Room-temperature ionic liquids. Solvents for synthesis and catalysis, *Chem. Rev.* **1999**, *99*, 2071-2084.

[18] N. V. Plechkova, K. R. Seddon, Applications of ionic liquids in the chemical industry, *Chem. Soc. Rev.* **2008**, *37*, 123-150.

[19] J. Hao, T. Zemb, Self-assembled structures and chemical reactions in room-temperature ionic liquids, *Curr. Opin. Colloid Interface Sci.* **2007**, *12*, 129-137.

[20] Z. Qiu, J. Texter; Ionic liquids in microemulsions, *Curr. Opin. Colloid Interface Sci.* **2008**, *13*, 252-262.

[21] T. L. Greaves, C. J. Drummond, Protic ionic liquids: Properties and applications, *Chem. Rev.* **2008**, *108*, 206-237.

[22] D. F. Evans, A. Yamauchi, R. Roman and E. Z. Casassa; Micelle formation in ethyl-ammonium nitrate, a low-melting fused salt, *J. Colloid Interface Sci.* **1982**, *88*, 89-96.

[23] D. F. Evans, S.-H. Chen, G. W. Schriver, E. M. Arnett, Thermodynamics of solution of nonpolar gases in a fused salt. Hydrophobic bonding behavior in a nonaqueous system, *J. Am. Chem. Soc.* **1981**, *103*, 481-482.

[24] D. F. Evans, A. Yamauchi, G. J. Wei, V. A. Bloomfield, Micelle size in ethylammonium nitrate as determined by classical and quasi-elastic light scattering. A, *J. Phys. Chem.* **1983**, *87*, 3537-3541.

[25] J. L. Anderson, V. Pino, E. C. Hagberg, V. V. Sheares, D. W. Armstrong, Surfactant solvation effects and micelle formation in ionic liquids, *Chem. Commun.* **2003**, 2444-2445.

[26] C. Patrascu, F. Gauffre, F. Nallet, R. Bordes, J. Oberdisse, N. de Lauth-Viguerie, C. Mingotaud, Micelles in ionic liquids: aggregation behavior of alkyl poly(ethyleneglycol)-ethers in 1-butyl-3-methyl-imidazolium type ionic liquids, *Chem. Phys. Chem.* **2006**, *7*, 99-101.

[27] J. Hao, A. Song, J. Wang, X. Chen, W. Zhuang, F. Shi, F. Zhou, W. Liu, Self-assembled structures in room-temperature ionic liquids, *Chem. Eur. J.* **2005**, *11*, 3936-3940.

[28] T. L. Greaves, A. Weerawardena, I. Krodkiewska, C. J. Drummond, Protic ionic liquids: Physicochemical properties and behavior as amphiphile self-assembly solvents, *J. Phys. Chem. B*, **2008**, *112*, 896-905.

[29] S. Thomaier, W. Kunz; Aggregates in mixtures of ionic liquids, *J. Mol. Liq.* **2007**, *130*, 104-107.

[30] S. Thomaier, Formulation and characterization of new innovative colloidal systems involving ionic liquids for the application at high temperatures, Ph.D. Thesis, University of Regensburg, 2009.

[31] Y. Zhao, X. Chen, X. Weng, Liquid crystalline phases self-organized from a surfactant-like ionic liquid C_{16}mimCl in ethylammonium nitrate, *J. Phys. Chem B*, **2009**, *113*, 2024-2030.

[32] P. Wasserscheid, T. Welton, Ionic liquids in syntheses, 2nd edition, ed. P. Wasserscheid, T. Welton, Wiley-VCH: Weinheim 2007.

[33] P. Wasserscheid, W. Keim, Ionic liquids – new "solutions" for transition metal catalysis, *Angew. Chem. Int. Ed.* **2000**, *39*, 3772-3789.

VII. Literature cited

[34] H. Weingaertner, Understanding ionic liquids at the molecular level: facts, problems, and controversies, *Angew. Chem. Int. Ed.* **2008**, *47*, 654-670.

[35] S. Gabriel, Ethylamine derivatives, Berichte der Deutschen Chemischen Gesellschaft **1888**, *21*, 566-575.

[36] T. L. Cottrell, J. E. Gill, Preparation and heats of combustion of some amine nitrates, *J. Chem. Soc.* **1951**, 1798-1800.

[37] I. Krossing, J. M. Slattery, C. Daguenet, P. J. Dyson, A. Oleinikova, H. Weingärtner, Why are ionic liquids liquid? A simple explanation based on lattice and solvation energies, *J. Am. Chem. Soc.* **2006**, *128*, 13427-13434.

[38] H. Xue, R. Verma, J. M. Shreeve, Review of ionic liquids with fluorine-containing anions, *J. Fluorine Chem.*, **2006**, *127*, 159-176.

[39] J. G. Huddleston, A. E. Visser, M. W. Reichert, H. D. Willauer, G. A. Broker, R. D. Rogers, Characterization and comparison of hydrophilic and hydrophobic room temperature ionic liquids incorporating the imidazolium cation, *Green Chem.* **2001**, *3*, 156-164.

[40] P. Bonhôte, A.-P. Dias, N. Papageorgiou, K. Kalyanasundaram, M. Grätzel, Hydrophobic, highly conductive ambient-temperature molten salts, *Inorg. Chem.* **1996**, *35*, 1168-1178.

[41] C. F. Poole, Chromatographic and spectroscopic methods for the determination of solvent properties of room temperature ionic liquids, *J. Chromatogr. A* **2004**, *1037*, 49–82

[42] A. Arfan, J. P. Bazureau, Efficient combination of recyclable task specific ionic liquid and microwave dielectric heating for the synthesis of lipophilic esters, *Org. Process Res. Dev.* **2005**, *9*, 743-748.

[43] D. Fang, X.-L. Zhou, Z.-W. Ye, Z.-L. Liu, brønsted acidic ionic liquids and their use as dual solvent-catalysts for Fischer esterifications, *Ind. Eng. Chem. Res.* **2006**, *45*, 7982-7984.

[44] D. R. MacFarlane, J. M. Pringle, K. M. Johansson, S. A. Forsyth, M. Forsyth, Lewis base ionic liquids, *Chem. Commun.* **2006**, 1905-1917.

[45] D. R. MacFarlane, K. R. Seddon, Ionic liquids - progress on the fundamental issues, *Aust. J. Chem.* **2007**, *60*, 3-5.

VII. Literature cited

[46] K. R. Seddon, Ionic liquids for clean technology, *J. Chem. Tech. Biotechnol.* **1997**, *68*, 351-356.

[47] Y. Chauvin, L. Mussmann, H. Olivier, A novel class of versatile solvents for two-phase catalysis: hydrogenation, isomerization, and hydroformylation of alkenes catalyzed by rhodium complexes in liquid 1,3-dialkylimidazolium salts, *Angew. Chem. Int. Ed.* **1996**, *34*, 2698-2700.

[48] D. Appleby, C. L. Hussey, K. R. Seddon, J. R. Turp, Room-temperature ionic liquids as solvents for electronic absorption spectroscopy of halide complexes, *Nature* **1986**, *323*, 614-616.

[49] M. J. Earle, J. M. S. S. Esperanca, M. A. Gilea, J. N. Canongia Lopes, L. P. N. Rebelo, J. W. Magee, K. R. Seddon, J. A. Widegren, The distillation and volatility of ionic liquids, *Nature* **2006**, *439*, 831-835.

[50] P. Wasserscheid, Volatile times for ionic liquids, *Nature* **2006**, *439*, 797.

[51] R. Ludwig, U. Kragl, Do we understand the volatility of ionic ligands?, *Angew. Chem.Int. Ed.* **2007**, *46*, 6582-6584.

[52] C. A. Angell, Formation of glasses from liquids and biopolymers, *Science,* **1995**, *267*, 1924-1935.

[53] P. Walden, Organic solvents and ionization media. III. Interior friction and its relation to conductivity, *Z. Phys. Chem.* **1906**, *55*, 207-249.

[54] W. Xu, C. A. Angell, Solvent-free electrolytes with aqueous solution-like conductivities, *Science* **2003**, *302*, 422-425.

[55] M. Yoshizawa, W. Xu, C. A. Angell, Ionic liquids by proton transfer: vapor pressure, conductivity, and the relevance ΔpK_a from aqueous solutions. *J. Am. Chem. Soc.* **2003**, *125*, 15411-15419.

[56] W. Xu, E. I. Cooper, C. A. Angell, Ionic Liquids: ion moblilities, glass temperatures, and fragilities. *J. Phys. Chem. B,* **2003**, *107*, 6170-6178.

VII. Literature cited

[57] K. J. Fraser, E. I. Izgorodina, M. Forsyth, J. L. Scott, D. R. MacFarlane, Liquids between "molecular" and "ionic" liquids: liquid ion pairs? *Chem. Commun.* **2007**, *37*, 3817-3819.

[58] T. L. Greaves, A. Weerawardena, C. Fong, I. Krodkiewska, C. J. Drummond, Protic ionic liquids: Solvents with tunable phase behavior and physicochemical properties, *J. Phys Chem B* **2006**, *110*, 22479-22487.

[59] C. Reichardt, Empirical parameters of solvent polarity, *Angew. Chem.*, **1965**, *77*, 30–40.

[60] M. J. Muldoon, C. M. Gordon and I. R. Dunkin; Investigations of solvent-solute interactions in room temperature ionic liquids using solvatochromic dyes. *J. Chem. Soc.* **2001**, *4*, 433-435.

[61] C. Reichardt, Polarity of ionic liquids determined empirically by means of solvatochromic pyridinium N-phenolte betaine dyes, *Green Chem.* **2005**, *7*, 339-351.

[62] C. Reichardt, E. Harbusch-Goernert, Pyridinium N-phenoxide betaines and their application for the characterization of solvent polarities. X. Extension, correction, and new definition of the ET solvent polarity scale by application of a lipophilic penta-tert-butyl-substituted pyridinium N-phenoxide betaine dye, *Liebigs Ann. Chem.* **1983**, *5*, 721–743.

[63] M. Maase, K. Massone, K. Halbritter, R. Noe, M. Bartsch, W. Siegel, V. Stegmann, M. Flores, O. Huttenloch, M. Becker, Methods for the separation of acids from chemical reaction mixtures by means of ionic fluids, WO 2003 062171.

[64] B. Weyershausen, K. Hell, U. Hesse, Industrial application of ionic liquids as performance additives, *Green. Chem.* **2005**, *7*, 283-287.

[65] T. P. Hoar, J. H. Schulman, Transparent water-in-oil dispersions: the oleophatic hydro-micelle, *Nature*, **1943**, *152*, 102-103.

[66] J. H. Schulman, W. Stoeckenius, L. M. Prince, Mechanism of formation and structure of micro emulsions by electron microscopy, *J. Phys. Chem.* **1959**, *63*, 1677-1680.

[67] I. Danielsson, B. Lindmann, The definition of microemulsions, *Colloids Surf.* **1981**, *3*, 391-392.

[68] P. A. Winsor, Solvent properties of amphiphilc compounds, Butterworth: London, 1954.

[69] J.-L. Salager, R. E. Antón, Handbook of Microemulsions Science and Technology, Ionic microemulsions, ed. P. Kumar, K. L. Mittal, Dekker: New York, 1999, chap. 8, 247-280.

[70] S. E. Friberg, M. Podzimek, A non-aqueous microemulsion, *Colloid Polym. Sci.* **1984**, *262*, 252-253.

[71] I. Rico, A. Lattes, "Anomalous" effect of a double bond on the micellization of unsaturated amines, *Nouv. J. Chim.* **1984**, *8*, 499-500.

[72] S. E. Friberg, Y.-C. Liang, Phase equilibria in the glycerol-aerosol OT system with decanol or hydrocarbon, *Surfactant Sci. Ser.* **1987**, *24*, 103-113.

[73] I. Rico, A. Lattes, Formamide as a water substitute. IX. Waterless microemulsions. 6. A new type of water-insoluble surfactants and nonaqueous microemulsions, *Surfactant Sci. Ser.* **1987**, *24*, 357-375.

[74] Z. Zaidi, C. Mathew, J. Peyrelasse, C. Boned, Percolation and critical exponents for the percolation of microemulsions, *Phys. Rev. A* **1990**, *42*, 872-876.

[75] S. Ray, S. P. Moulik, Dynamics and thermodynamics of Aerosol OT-aided nonaqueous microemulsions, *Langmuir*, **1994**, *10*, 2511-2515.

[76] B. K. Paul, S. P. Moulik, Microemulsions: An overview, *J. Dispersion Sci. Technol.* **1997**, *18*, 301-367.

[77] L. M. Prince (ed.), Microemulsions: Theory and practise, Academic Press: New York, 1977.

[78] T. Hellweg, Phase structures of microemulsions, *Current Opin. Colloid Interface Sci.* **2002**, *7*, 50-56.

[79] R. P. Bagwe, K. C. Khilar, Effects of the intermicellar exchange and cations on the size of silver chloride nanoparticles formed in reverse micelles of AOT, *Langmuir* **1997**, *13*, 6432-6438.

[80] R. P. Bagwe, K. C. Khilar, Effects of intermicellar exchange rate on the formation of silver nanoparticles in reverse microemulsions of AOT, *Langmuir* **2000**, *16*, 905-910.

VII. Literature cited

[81] P. Pieruschka, S. Marcelija, Monte-Carlo simulations of curvature elastic interfaces, *Langmuir* **1994**, *10*, 345-350.

[82] L. E. Scriven, Equilibrium bicontinuous structure, *Nature* **1976**, *263*, 123-125.

[83] M. Mihailescu M, Monkenbusch, H. Endo, J. Allgaier, G. Gompper, J. Stellbrink, D. Richter, B. Jakobs, T. Sottmann, B. Farago, Dynamics of bicontinuous microemulsion phases with and without amphiphilic block-co-polymers, *J. Chem. Phys.* **2001**, *115*, 9563-9577.

[84] M. Maugey, A. M. Bellocq, Effect of adsorbed and anchored polymers on membrane flexibility: a light scattering study of sponge phases, *Langmuir* **2001**, *17*, 6740-6742.

[85] M. Dubois, T. Zemb, Swelling limits for bilayer microstructures: the implosion of lamellar structure vs. disordered lamellae, *Curr. Opin. Coll. Interface Sci.* **2000**, *5*, 27-37.

[86] P. A. Winsor, Hydrotropy, solubilisation and related emulsification processes. PartI. *Trans. Farady. Soc.* **1948**, *44*, 376- 382.

[87] A. M. Bellocq, J. Biais, P. Bothorel, B. Clin, G. Fourche, P. Lalanne, B. Lemaire, B. Lemanceau, D. Roux, Microemulsions, *Adv. Colloid Interface Sci.* **1984**, *20*, 167- 272.

[88] D. F. Evans, H. Wennerström, The colloidal domain, second edition, Wiley: New York, 1999.

[89] J. M. Schick, Nonionic Surfactants, Surfactant Science Series, Dekker: New York, 1987, 23, 753-833.

[90] K. Wormuth, O. Lade, M. Lade, R. Schomäcker, Microemuslions, Handbook of Applied Surface and Colloid Chemistry, ed. K. Holmberg, Wiley: U.K. 2002, 2, 55-77.

[91] M. S. Leaver, U. Olsson, H. Wennerström, Phase behavior and structure in a non-ionic surfactant-oil-water mixture, *J. Chem. Soc. Faraday Trans.* **1995**, *91*, 4269-4274.

[92] M. Kahlweit, R. Strey; D. Haase, H. Kunieda, T. Schmeling, B. Faulhaber, M. Borcovec, H. F. Eicke, G. Busse, F. Eggers, Th. Funck, H. Richmann, L. Magid, O. Söderman, P. Stilbs, J. Winkler, A. Dittrich, W. Jahn, How to study microemulsions, *J. Colloid Interface Sci.* **1987**, *118*, 436-453.

[93] U. Olsson, H. Wennerström, Globular and bicontinuous phases of nonionic surfactant films, *Adv. Colloid Interface Sci.* **1994**, *49*, 113-146.

[94] S. Saeki, N. Kuwahara, M. Nakata, M. Kaneko, Upper and lower critical solution temperatures in poly(ethylene glycol) solutions, *Polymer* **1976**, *17*, 685-689.

[95] G. Karlström, A new model for upper and lower critical solution temperatures in poly(ethylene oxide) solutions, *J. Phys. Chem.*, **1985**, *89*, 4962-4964.

[96] M. Kahlweit, R. Strey, P. Firman, D. Haase, J. Jen, R. Schomäcker, General patterns of the phase behavior of mixtures of H_2O, nonpolar solvents, amphiphiles, and electrolytes, *Langmuir* **1988**, *4*, 499-511.

[97] M. Kahlweit, R. Strey, R. Schomäcker, D. Haase, General patterns of the phase behavior of mixtures of H_2O, nonpolar solvents, amphiphiles, and electrolytes. 2, *Langmuir* **1989**, *5*, 305-315.

[98] M. Kahlweit, R. Strey, Phase behavior of ternary systems of the type H_2O-oil-nonionic amphiphile (microemulsions), *Angew. Chem. Int. Ed.* **1985**, *24*, 654-668.

[99] S. I. Ahmad, K. Shinoda, S. Friberg, Microemulsions and phase equilibria, *J. Colloid Interface Sci.* **1974**, *47*, 32-37.

[100] P. Ekwall, Composition, properties, and structures of liquid crystalline phases in systems of amphiphilic compounds, *Adv. Liq. Cryst.* **1975**, *1*, 1-142.

[101] M. Zulauf, H.-F. Eicke, Inverted micelles and microemulsions in the ternary system H_2O/Aerosol-OT/isooctane as studied by photon correlation spectroscopy, *J. Phys. Chem.* **1979**, *83*, 480-486.

[102] H.-F. Eicke, Z. Markovic, Temperature dependent coalescence in water-oil microemulsions and phase transitions to lyotropic mesophases, *J. Colloid Interface Sci.* **1981**, *79*, 151-159.

[103] S. J. Chen, D. F. Evans, B. W. Ninham, D. J. Mitchell, F. D. Blum, S. Pickups, Curvature as a determinant of microstructure and microemulsions, *J. Phys. Chem.* **1986**, *90*, 842-847.

VII. Literature cited

[104] V. Chen, G. G. Warr, D. F. Evans, Curvature and geometric constraints as determinants of microemulslon structure: Evidence from fluorescence anisotropy measurements, *J. Phys. Chem.* **1988**, *92*, 768-773.

[105] K. Fontell, A. Ceglie, B. Lindman, B. Ninham, Some observations on phase diagrams and structure in binary and ternary systems of didodecyldimetylammonium bromide, *Acta Chem. Scand.* **1986**, *A40*, 247-256.

[106] D. J. Mitchell, B. W. Ninham, Micelles, vesicles and microemulsions, *J. Chem. Soc., Faraday Trams. 2* **1981**, *77*, 601-629.

[107] A. K. Rakshit, S. P. Moulik, Physicochemistry of w/o microemulsions: Formation, stability and droplet clustering, ed. M. Fanun, Surfactant Science Series, Dekker: New York, 2009, 144, 17-57.

[108] I. Carlsson, A. Fogden, H. Wennerström, Electrostatic interactions in ionic microemulsions, *Langmuir*, **1999**, *15*, 6150-6155.

[109] K. R. Wormuth, E. W. Kaler, Amines as microemsulsion cosurfactants, *J. Phys. Chem.* **1987**, *91*, 611-617.

[110] M. Bourrel, R. S. Schechter, Microemulsions and related systems, ed. M. Bourrel, R. S. Schechter, Surfactant Science Series, Dekker: New York, 1988, 30, 127-278.

[111] G. D. Rees, B. H. Robinson, Microemulsions and oragnogels: properties and applications, *Adv. Mater.* **1993**, *5*, 608-619.

[112] B. K. Paul, S. P. Moulik, Uses and applications of microemulsions, *Current Sci.* **2001**, *80*, 990-1001.

[113] M. Fanun, Microemulsions properties and applications, ed. M. Fanun, Surfactant Science Series, Dekker : New York, 2009, 144.

[114] D. O. Shah, Surface phenomena in enhanced oil recovery, Plenum: New York, 1981.

[115] D. O. Shah, Micelles, microemulsions and monolayers: Science and technology, ed. D. O. Shah, Dekker: New York, 1998.

[116] M. J. Schwuger, K. Stickdorn, R. Schömäcker, Microemulsions in technical processes, *Chem. Rev.* **1995**, *95*, 849-864.

[117] K. Bonkoff, M. J. Schwuger, G. Subklew, Use of microemulsions fort he extraction of contaminated solids, Industrial applications of microemulsions, ed. C. Solans, H. Kunieda, Marcel Dekker, New York, 1997, pp.355.

[118] C. Solans, H. Kunieda, Industrial applications of microemulsions, ed. C. Solans, H. Kunieda, Surfactant Science Series, Dekker: New York, 1997,66.

[119] J. Eastoe, B. Warne, Nanoparticle and polymer synthesis in microemulsions, *Curr. Opinion Colloid Interface Sci.* **1996**, *1*, 800-805.

[120] A. P. Full, J. E. Puig, L. U. Gron, E. W. Kaler, J. R. Minter, T. H. Mourey, J. Texter, Polymerization of tetrahydrofurfuryl methacrylate in three-component anionic microemulsions, *Macromolecules*, **1992**, *25*, 5157-5164.

[121] J. Kemken, A. Ziegler, B. W. Mueller, Investigations into the pharmacodynamic effects of dermally administered microemulsions containing β-blockers *J. Pharm. Pharmacol.* **1991**, *43*, 679-684.

[122] J. Carlfors, I. Blute, V. Schmidt, Lidocaine in microemulsion - a dermal delivery system, *J. Disp. Sci. Technol.* **1991**, *12*, 467-482.

[123] K. Shinoda, Y. Shibata, B. Lindmann, Interfacial Tensions for Lecithin Microemulsions Including the Effect of Surfactant and Polymer Addition, *Langmuir*, **1993**, *9*, 1254-1257.

[124] A. Gomez-Puyon, (ed.), Biomol. Org. Solvents, CRC Press, Boca Raton, 1992.

[125] M. A. Biasutti, E. B. Abuin, J. J. Silber, N. M. Correa, E. A. Lissi, Kinetics of reactions catalyzed by enzymes in solutions of surfactants, *Adv. Colloid Interface Sci.* **2008**, *136*, 1-24.

[126] W. Kunz, D. Touraud, P. Bauduin, Enzyme kinetics as useful probe for micelle and microemulsion structure and dynamics, ed. M. Fanun, *Surfactant Science Ser.* **2009**, *144*, 331-347.

VII. Literature cited

[127] M. Lagues, R. Ober, C. Taupin, Study of structure and electrical conductivity in microemulsions: evidence for percolation mechanism and phase inversion, *J. Phys. Lett.* **1978**, *39*, L487-L491.

[128] H.-F. Eicke, M. Borkovec, B. Das-Gupta, Conductivity of water-in-oil microemulsions: a quantitative charge fluctuation model, *J. Phys. Chem.* **1989**, *93*, 314-317.

[129] H. B. Callen, Thermodynamics, Wiley, New York, 1960.

[130] N. Kallay, A. Chittofrati, Conductivity of microemulsions: refinement of charge fluctuation model, *J. Phys. Chem.* **1990**, *94*, 4755-4756.

[131] N. Kallay, M. Tomic, A. Chittofrati, Conductivity of water-in-oil microemulsions: comparison of the Boltzmann statistics and the charge fluctuation model, *Colloid Polym. Sci.* **1992**, *270*, 194-196.

[132] D. G. Hall, Conductivity of microemulsions: An improved charge fluctuation model, *J. Phys. Chem.* **1990**, *94*, 429-430.

[133] B. Halle, M. Björling, Microemulsions as macroelectrolytes, *J. Chem. Phys.* **1995**, *103*, 1655-1668.

[134] G. Grest, I. Webmann, S. Safran, A.Bug, Dynamic percolation in microemulsion, *Phys. Rev. A* **1986**, *33*, 2842-2845.

[135] J. Peyrelasse, M. Moha-Ouchane, C. Boned, Dielectric relaxation and percolation phenomena in ternary microemulsions, *Phys. Rev. A* **1988**, *A38*, 904-917.

[136] A. Jada, J. Lang, R. Zana, Relation between electrical percolation and rate constant for exchange of material between droplets in water in oil microemulsions, *J. Phys. Chem.* **1989**, *93*, 10-12.

[137] A. Jada, J. Lang, R. Zana, Ternary water in oil microemulsions made of cationic surfactants, water, and aromatic solvents. 1. Water solubility studies, *J. Phys. Chem.* **1990**, *94*, 381-387.

[138] M. Lagues, Electrical conductivity of microemulsions: a case of stirred percolation, *J. Physique Lett.* **1979**, *40*, L331-L333.

[139] A. Molski, E. Dutkiewicz, Electrical conductivity and percolation in water-in-oil microemulsions, *Pol. J. Chem.* **1996**, *70*, 959-971.

[140] M. W. Kim, J. S. Huang, Percoaltionlike phenomena in oil-continuous microemulsions, *Phys. Rev. A* **1986**, *34*, 719-722.

[141] B. Derrida, D. Stauffer, H. J. Hermann, J. Vannimenus, Transfer matrix calculation of conductivity in three-dimensional random resistor networks at percolation threshold, *J. Physique Lett.* **1983**, *44*, L701-L706.

[142] S. P. Moulik, B. K. Paul, Structure, dynamics and transport properties of microemulsions, *Adv. Colloid Interface Sci.* **1998**, *78*, 99-195.

[143] S. Ajith, A. C. John, A. R. Rakshit, Physicochemical studies of microemulsions, *Pure Appl. Chem.* **1994**, *66*, 509-514.

[144] A. C. John, A. K. Rakshit, Phase behavior and properties of a microemulsion in the presence of NaCl, *Langmuir*, **1994**, *10*, 2084-2087.

[145] S. P. Moulik, A. K. Rakshit, Physicochemistry and applications of microemulsions, *J. Surf. Sci. Technol.* **2006**, *22*, 159-186.

[146] M. Borkovec, H. F. Eicke, H. Hammerich, B. Das Gupta, Two percolation processes in microemulsions, *J. Phys. Chem.* **1988**, *92*, 206-211.

[147] R. Finsey, Particle sizing by quasi-elastic light scattering, *Adv. Colloid Interface Sci.* **1994**, *52*, 79-143.

[148] S. Provencher, A constrained regularization method for inverting data represented by linear algebraic or integral equations, *Computer Phys. Comm.* **1982**, *27*, 213-227.

[149] S. Provencher, CONTIN: A general purpose constrained regularization program for inverting noisy linear algebraic and integral equations, *Computer Phys. Comm.* **1982**, *27*, 229-242.

[150] E. Gabrowski, I. Morrison, Particle size distributions from analysis of quasi-elastic light-scattering Data, chap. 7, measurements of suspended particles by quasielastic light scattering, ed. B. Dahneke, Wiley, New York, 1983.

VII. Literature cited

[151] M. Kotlarchyk, S. H. Chen, J. S: Huang, Temperature dependence of size and polydispersity in a three-component microemulsion by small-angle neutron scattering, *J. Phys. Chem.* **1982**, *86*, 3273-3276.

[152] D. R. Caudwell, J. P. M. Trusler, V. Vesovic, W. A. Wakeham, The viscosity and density of n-dodecane and *n*-octadecane at pressures up to 200 MPa and temperatures up to 473 K, Int. *J. Thermophys.*, **2004**, *25*, 1339-1352.

[153] A.-J. Dianouy, G. Lander, Neutron Data Booklet, 2nd edition, Institute Laue-Langevin, 2003.

[154] S. M. King, Small-angle neutron scattering, report published on ISIS Web page, 1997.

[155] O. Glatter, O. Kratky, *Small Angle X- ray Scattering*, ed. O. Glatter, O. Kratky, Academic Press, London, 1982, p. 119.

[156] P. N. Pusey, Neutrons, X-ray and light: scattering methods applied to soft matter, ed. P. Lindner, T. Zemb, North Holland, Amsterdam, chap. 1, 2002, 3-21.

[157] T. Zemb, Neutrons, X-ray and light: scattering methods applied to soft matter, ed. P. Lindner, T. Zemb, North Holland, Amsterdam, chap. 13, 2002, 317-350.

[158] R. Strey, O. Glatter, K.-V. Schubert, E. W. Kaler, Small-angle neutron scattering of D2O–C12E5 mixtures and microemulsions with n-octane: Direct analysis by Fourier transformation, *J. Chem. Phys.* **1996**, *105*, 1175-1188.

[159] F. Lichterfeld, T. Schmeling, R. Strey; Microstructure of microemulsions of the system water-n-tetradecane-alkyl polyglycol ether ($C_{12}E_5$), *J. Phys. Chem.* **1986**, *90*, 5762-5766.

[160] J. Eastoe, S. Gold, S. E. Rogers, A. Paul, T. Welton, R. Heenan, I. Grillo, Ionic liquid-in-oil microemulsions, *J. Am. Chem. Soc.* **2005**, *127*, 7302-7303.

[161] M. Teubner, R. Strey, Origin of the scattering peak in microemulsions, *J. Chem. Phys.* **1987**, *87*, 3195-3200.

[162] T. Shimobouji, H. Matasuoka, N. Ise, Small-angle x-ray scattering studies on nonionic microemulsions, *Phys. Rev. A* **1989**, *39*, 4125-4131.

[163] G. Porod, The X-ray small-angle scattering of close-packed colloid systems. I., Kolloid Z. **1951**, *124*, 83-114.

[164] P. Debye, A. M. Bueche, Scattering by an inhomogeneous solid, *J. Appl. Phys.* **1949**, *20*, 518-525.

[165] P. Debye, H. R. Anderson, H. Brumberger, Scattering by an inhomogeneous solid. II. The correlation function and its application, *J. Appl. Phys.* **1957**, *28*, 679-683.

[166] S. H. Chen, S. L. Chang, R. Strey, On the interpretation of scattering peaks from bicontinuous microemulsions, *Prog. Collopid Polymer Sci.* **1990**, *81*, 30-35.

[167] K.-V. Schubert, R. Strey, S. R. Kline, E. W. Kaler, Small angle neutron scattering near Lifshitz lines: transition from weakly structured mixtures to microemulsions, *J. Chem. Phys.* **1994**, *101*, 5343-5355.

[168] H. Leitao, M. M. T. da Gama, R. Strey, Scaling of the interfacial tension of microemulsions: a Landau theory approach, *J. Chem. Phys.* **1998**, *108*, 4189-4198.

[169] S. Engelskirchen, N. Elsner, T. Sottmann, R. Strey; Triacylglycerol microemulsions stabilized by alkyl ethoxylate surfactants-A basic study; phasebehavior, interfacial tension and microstructure, *J. Colloid Interface Sci.* **2007**, *312*, 114-121.

[170] O. Spalla, Neutrons, X-ray and light: Scattering methods applied to soft condensed matter, ed. P. Lindner, T. Zemb, T. North Holland, Amsterdam, chap. 3, 2002, 49-71.

[171] J. Jouffroy, P. Levinson, P. G. De Gennes, Phase equilibriums involving microemulsions. (Remarks on the Talmon - Prager model), *J. Phys. France* **1982**, *43*, 241-248.

[172] S. T. Milner, S. A. Safran, D. Andelman, M. E. Cates, D. Roux, Correlations and structure factor of bicontinuous microemulsions, *J. Phys. France* **1988**, *49*, 1065-1076.

[173] M.-P. Pileni, T. Zemb, C. Petit, Solubilization by reverse micelles: solute localization and structure perturbation, *Chem. Phys. Lett.* **1985**, *118*, 414-420.

[174] P. N. Pusey, H. M. Fijnaut, A. Vrij, Mode amplitudes in dynamic light scattering by concentrated liquid suspensions of polydisperse hard spheres, *J. Chem. Phys.* **1982**, *77*, 4270-4281.

VII. Literature cited

[175] W. L. Grffith, R. Triolo, A. L. Compere, Analytical structure function of a polydisperse Percus-Yevick fluid with Schulz (gamma) distributed diameters, *Phys. Rev. A* **1986**, *33*, 2197-2200.

[176] W. L. Grffith, R. Triolo, A. L. Compere, Analytical scattering function of a polydisperse Percus-Yevick fluid with Schulz- (Γ-) distributed diameters, *Phys. Rev. A* **1987**, *35*, 2200-2206.

[177] N. Freiberger, C. Moitzi, L. de Campo, O. Glatter, An attempt to detect bicontinuity from SANS data, *J. Colloid Interface Sci.* **2007**, *312*, 59-67.

[178] O. Glatter, A new method for the evaluation of small-angle scattering data, *J. Appl. Cryst.* **1977**, *10*, 415-421.

[179] J. Brunner-Popela, O. Glatter, Small-angle scattering of interacting particles. I. Basic principles of a global evaluation technique, *J. Appl. Cryst.* **1997**, *30*, 431-442.

[180] A. Bergmann, G. Fritz, O. Glatter, Solving the generalized indirect Fourier transformation (GIFT) by Boltzmann simplex simulated annealing (BSSA), *J. Appl. Cryst.* **2000**, *33*, 1212-1216.

[181] O. Glatter, D. Orthaber, A. Stradner, G. Scherf, M. Fanun, N. Garti, V. Clément, M. E. Leser, Sugar-ester nonionic microemulsion: Structural characterization, *J. Colloid Interface Sci.* **2001**, *241*, 215-225.

[182] R. Strey, Microemulsion microstructure and interfacial curvature, *Colloid Polym. Sci.* **1994**, *272*, 1005-1019.

[183] W. Jahn, R. Strey, Microstructure of microemulsions by freeze fracture electron microscopy, *J. Phys. Chem.* **1988**, *92*, 2294-2301.

[184] P. K. Vinson, J. G. Sheehan, W. G. Miller, L. E. Scriven, H. T. Davis, Viewing microemulsions with freeze-fracture transmission electron microscopy, *J. Phys. Chem.* **1991**, *95*, 2546-2550.

[185] L. Belkoura, C. Stubenrauch, R. Strey, Freeze fracture direct imaging: A new freeze fracture method for specimen preparation in cry-transmission electron microscopy, *Langmuir* **2004**, *20*, 4391-4399.

[186] K. Shinoda, B. Lindman, Organized surfactant systems: Microemulsions, *Langmuir*, **1987**, *3*, 135-149.

[187] U. Olsson, K. Shinoda, B. Lindman, Change in structure of microemulsions with the hydrophile-lipophile balance of nonionic surfactant as revealed by NMR self-diffusion studies, *J. Phys. Chem.* **1986**, *90*, 4083-4088.

[188] O. Söderman, N. Nydén, NMR in microemulsions. NMR translational diffusion studies of a model microemulsion, *Colloids Surf. A* **1999**, *158*, 273-280.

[189] U. Olsson, Handbook of Applied Surface and Colloid Chemsitry, ed. K. Holmberg; John Wiley & Sons Ltd.; Chichester, U.K., 2002, vol. 2, 333-356.

[190] O. Rojas, J. Koetz, S. Kosmella, B. Tiersch, P. Wacker, M. Kramer, Structural studies of ionic liquid-modified microemulsions, *J. Colloid Interface Sci.* **2009**, *333*, 782-790.

[191] B. Lindman, U. Olsson, Structure of microemulsions studied by NMR, *Ber. Bunsenges. Phys. Chem.* **1996**, *100*, 344-363.

[192] R. Atkin, G. G. Warr; Phase behavior and microstructure of microemulsions with a room-temperature ionic liquid as the polar phase. *J. Phys. Chem. B* **2007**, *111*, 9309-9316.

[193] J. D. Holbrey, K. R. Seddon, The phase behavior of 1-alkyl-3-methylimidazolium tetrafluoroborates; ionic liquids and ionic liquid crystals, *J. Chem. Soc., Dalton Trans.* **1999**, *13*, 2133-2140.

[194] L. Cammarata, S. G. Kazarian, P. A. Salter, T. Welton, Molecular states of water in room temperature ionic liquids. *Phys. Chem. Chem. Phys.* **2001**, *3*, 5192-5200.

[195] O. Zech, S. Thomaier, P. Bauduin, T. Rueck, D. Touraud, W. Kunz, Microemulsions with an ionic liquid surfactant and room temperature ionic liquids as polar pseudo-phase. *J. Phys. Chem. B* **2009**, *113*, 465-473.

[196] J. D. Holbrey, W. M. Reichert, R. P. Swatloski, G. A. Broker, W. R. Pitner, K. R. Seddon, R. Rogers, Efficient, halide free synthesis of new, low cost ionic liquids: 1,3-dialkylimidazolium salts containing methyl- and ethyl-sulfate anions, *Green Chem.* **2002**, *4*, 407-413.

VII. Literature cited

[197] K. Matsushima, N. Kawamura, M. Okahara, Synthesis of novel macrocyclic ether-ester compounds via the intramolecular cyclization of oligoethylene glycol monocarboxymethyl ethers, *Tetrahedron Lett.* **1979**, *20*, 3445-3448.

[198] J. Barthel, R. Wachter, H.-J. Gores, *Modern Aspects of Electrochemistry*, Ed. B. E. Conway, J. O´M Bockris, 1979, 13, 1-79.

[199] R. Wachter, J. Barthel, Untersuchungen zur Temperaturabhängigkeit der Eigenschaften von Elektrolytlösungen II. Bestimmung der Leitfähigkeit über einen großen Temperaturbereich. *Ber. Bunsenges. Phys. Chem* **1979**, *83*, 634-642.

[200] J. Barthel, F. Feuerlein, R. Neueder, R. Wachter, Calibration of conductance cells at various temperatures. *J. Sol. Chem.* **1980**, *9*, 209-219.

[201] R. A. Robinson, R. H. Stokes, *Electrolyte Solutions*, 2nd edn. Butterworth, London, 1970.

[202] D. R. Lide, CRC Handbook of Chemistry and Physics, CRC Press: Boca Raton, FL, 2004.

[203] T. Zemb, O. Tache, F. Ne, O. Spalla; A high sensitivity pinhole camera for soft condensed matter, *J. Appl. Cryst.* **2003**, *36*, 800-805.

[204] T. Zemb, O. Tache, F. Ne, O. Spalla; Improving sensitivity of a small angle x-ray scattering camera with pinhole collimation using separated optical elements, *Rev. Sci. Instrum.* **2003**, *74*, 2456-2462.

[205] F. Ne, A. Gabriel, M. Kocsis, T. Zemb, Smearing effects introduced by the response function of position-sensitive gas detectors in SAXS experiments, *J. Appl. Cryst.* **1997**, *30*, 306-311.

[206] Y. Li, R. Beck, T. Huang, Tuo; M. C. Choi, M. Divinagracia,, Scatterless hybrid metal-single-crystal slit for small-angle X-ray scattering and high-resolution X-ray diffraction, *J. Appl. Cryst.* **2008**, *41*, 1134-1139.

[207] I. Grillo, *Effect of instrumental resolution and polydispersity on ideal form factor in small-angle neutron scattering*. ILL Technical Report **2001**, *ILL01GR08T*, 1- 20.

[208] http://www.ill.eu/d22/

[209] I. Grillo, Soft matter characterization, ed.. R. Borsali, R. Pecora, Springer: Berlin, 2008; chap. 13, 725-782.

[210] J. S. Pedersen, D. Posselt, K. Mortensen, Analytical treatment of the resolution function for small-angle scattering, *J. Appl. Cryst.* **1990**, 23, 321-333.

[211] P. A. Z. Suarez, S. Einloft, J. E. L: Dullius, R. F. de Souza, J. Dupont, Synthesis and physical-chemical properties of ionic liquids based on l-butyl-3-methylimidazolium cation, *J. Chim. Phys.* **1998**, *95*, 1626-1639.

[212] H. Tokuda, K. Hayamizu, K. Ishii, M. A. B. H. Susan, M. Watanabe, Physicochemical properties and structures of room temperature ionic liquids. 1. Variation of anionic species, *J. Phys. Chem. B*, **2004**, *108*, 16593-16600.

[213] J. A. Widegren, E. M. Saurer, K. N. Marsh, J. W. Magee, Electrolytic conductivity of four imidazolium-based room-temperature ionic liquids and the effect of a water impurity, *J. Chem. Thermodyn.*, **2005**, *37*, 569-575.

[214] H. Tokuda, S. Tsuzuki, M. A. B. H. Susan, K. Hayamizu, M. Watanabe, How ionic are room-temperature ionic liquids? An indicator of the physicochemical properties, *J. Phys. Chem. B* **2006**, *110*, 19593-19600.

[215] Y. Yoshida, O. Baba, G. Saito, Ionic liquids based on dicyanamide anion: influence of structural variations in cationic structures on ionic conductivity, *J. Phys. Chem. B.* **2007**, *111*, 4724-4749.

[216] R. Ge, C. Hardacre, P. Nancarrow, D. W. Rooney, Thermal conductivities of ionic liquids over the temperature range from 293 K to 353 K, *J. Chem. Eng. Data* **2007**, *52*, 1819-1823.

[217] Y. O. Andryko, W. Reischl, G. E. Nauer, Trialkyl-substituted imidazolium-based ionic liquids for electrochemical applications: basic physicochemical properties, *J. Chem. Eng. Data* **2009**, *54*, 855-860.

[218] A. Stoppa, J. Hunger, R. Buchner, Conductivities of binary mixtures of ionic liquids with polar solvents, *J. Chem. Eng. Data* **2009**, *54*, 472-479.

[219] J. Hunger, A. Stoppa, S. Schrödle, G. Hefter, R. Buchner, Temperature dependence of the dielectric properties and dynamics of ionic liquids, *ChemPhysChem* **2009**, *10*, 723-733.

[220] C. P. Fredlake, J. M. Crosthwaite, D. G. Hert, N. V. K. Aki, J. F. Brennecke, Thermophysical properties of imidazolium-based ionic liquids, *J. Chem. Eng. Data* **2004**, *49*, 954-964.

[221] S. V. Dzyuba, R. A: Bartsch, Influence of structural variations in 1-alkyl(aralkyl)-3-methylimidazolium hexafluorophosphates and bis(trifluoromethylsulfonyl) imides on physical properties of the ionic liquids. ChemPhysChem 2002, 3, 161-166.

[222] A. Heintz, D. Klasen, J. K. Lehmann, C. Wertz, Excess molar volumes and liquid–liquid equilibria of the ionic liquid 1-methyl-3-octylimidazolium tetrafluoroborate mixed with butan-1-ol and pentan-1-ol. *J. Solution Chem.* **2005**, *34*, 1135-1144.

[223] G.-H. Tao, L. He, W. Liu, L. Xu, W. Xiong, T. Wang, Y. Kou, Preparation, characterization and application of amino acid-based green ionic liquids, *Green Chem.* **2006**, *8*, 639-.646

[224] D. F. Evans, E. W. Kaler, W. J. Benton; Liquid crystals in a fused salt: distearoyl-phosphatidylcholine in *N*-ethylammonium nitrate *J. Phys Chem.* **1983**, *87*, 533-535.

[225] K. Fumino, A. Wulf, R. Ludwig, Hydrogen bonding in protic ionic liquids: reminiscent of water, *Angew. Chem. Int. Ed.* **2009**, *48*, 3184-3286.

[226] R. Ludwig, A simple explanation for the occurrence of specific large aggregated ions in some protic ionic liquids, *J. Phys. Chem. B*, in press.

[227] T. L. Greaves, C. J. Drummond, Ionic liquids as amphiphile self-assembly media, *Chem. Soc. Rev.* **2008**, *37,* 1709-1726.

[228] M. U. Araos, G. G. Warr; Self-assembly of nonionic surfactants into lyotropic liquid crystals in ethylammonium nitrate, a room-temperature ionic liquid *J. Phys. Chem. B,* **2005**, *109*, 14275-14277.

[229] J. Bowers, C. P. Butts, P. J. Martin, M. C. Vergara-Gutierrez, R. K. Heenan, Aggregation behavior of aqueous solutions of ionic liquids, *Langmuir* **2004**, *20*, 2191-2198.

[230] Z. Miskolczy, K. Sebok-Nagy, L. Biczok, S. Goektuerk; Aggregation and micelle formation of ionic liquids in aqueous solution. *Chem. Phys. Lett.* **2004**, *400*, 296-300.

[231] J. Sirieix-Plenet, L. Gaillon, P. Letellier, Behavior of a binary solvent mixture constituted by an amphiphilic ionic liquid, 1-decyl-3-methylimidazolium bromide and water. Potentiometric and conductimetric studies, *Talanta* **2004**, *63*, 979-986.

[232] H. Kaper, B. Smarsly, Templating and phase behavior of the long chain ionic liquid C_{16}mimCl, *Z. Phys. Chem.* **2006**, *220*, 1455-1471.

[233] O. A. El Seoud, P. A. R. Pires, T. Abdel-Moghny, E. L. Bastos; Synthesis and micellar properties of surface-active ionic liquids: 1-alkyl-3-methylimidazolium chlorides *J. Colloid Interface Sci.* **2007**, *313*, 296-304.

[234] D. R. Chang, Conductivity of molten salts in the presence of oil and surfactant, *Langmuir* **1990**, *6*, 1132-1135.

[235] H. Gao, J. Li, B. Han, W. Chen, J. Zhang, R. Zhang, D. Yan, Microemulsions with ionic liquid polar domains. *Phys. Chem. Chem. Phys.* **2004**, *6*, 2914-2916.

[236] J. Li, J. Zhang, H. Gao, B. Han, L. Gao, Nonaqueous microemulsions containing ionic liquid [bmim][PF6] as polar microenvironment, *Colloid Polym. Sci.* **2005**, 283, 1371-1375.

[237] Y. Gao, J. Zhang, H. Xu, X. Zhao, L. Zheng, X. Li, L. I. Yu, Structural studies of 1-butyl-3-methylimidazolium tetrafluoroborate/TX-100/p-xylene ionic liquid microemulsions, *Chem. Phys. Chem.* **2006**, *7*, 1554-1561.

[238] N. Li, Y. Gao, L. Zheng, J. Zhang, L. Yu, X. Li, Studies on the micropolarities of bmimBF4/TX-100/Toluene ionic liquid microemulsions and their behaviors characterized by UV-visible spectroscopy, *Langmuir* 2007, *23*, 1091-1097.

[239] N. A. Li, S. Zhang, L. Zheng, Y. Gao, L. I. Yu, Second virial coefficient of bmimBF4/Triton X-100/cyclohexane ionic liquid microemulsion as investigated by microcalorimetry, *Langmuir* **2008**, *24*, 2973-2976.

[240] M. Clausse, L. Nicolas-Morgantini, A. Zradba, D. Touraud, Microemulsion systems, ed. H. L. Rosano, M. Clausse, Surfactant Science Series, Dekker: New York, 1987, 24, 15-298.

[241] M. Tomsic, M. Bester-Rogac, A. Jamnic, W. Kunz, D. Touraud, A. Bergmann, O. Glatter, Ternary systems of nonionic surfactant Brij 35, water and various simple alcohols: structural

investigations by small-angle X-ray scattering and dynamic light scattering, *J. Colloid Interface Sci.* **2006**, *294*, 194-211.

[242] T. Tlusty, S. A. Safran, R. Strey, Topology, phase instabilities, and wetting of microemulsion networks, *Phys. Rev. Lett.* **2000**, *84*, 1244-1247.

[243] B. Lagourette, J. Peyrelasse, C. Boned, M. Clausse, Percolative conduction in microemulsion type systems, Nature **1979**, *281*, 60-62.

[244] H. F. Eicke, A. Denss, Solution Chemistry of surfactants, Plenum: New York, 1979.

[245] C. Blattner, J. Bittner, G. Schmeer, W. Kunz; Electrical conductivity of reverse micelles in supercritical carbon dioxide, *Phys. Chem. Chem. Phys.* **2002**, *4*, 1921-1927.

[246] S. Schroedle, R. Buchner, W. Kunz; Percolating microemulsions of nonionic surfactants probed by dielectric spectroscopy. *ChemPhysChem* **2005**, *6*, 1051-1055.

[247] G. Porod, Small Angle X-ray Scattering, ed. O. Glatter, O. Kratky, Academic Press: London, 1982, chap. 2.

[248] J. Reimer, O. Söderman, T. Sottmann, K. Kluge, R. Strey, Microstructure of alkyl glucoside microemulsions: control of curvature by interfacial composition, *Langmuir* 2003, *19*, 10692-10702.

[249] S.-H. Chen, S.-L. Chang, R. Strey, Structural evolution within the one-phase region of a three-component microemulsion system: water-n-decane-sodium bisethylhexylsulfosuccinate (AOT), *J. Phys. Chem.* **1990**, *93*, 1907-1918.

[250] G. Körösl, E. S. Kováts, Density and surface tension of 83 organic liquids, *J. Chem. Eng. Data* **1981**, *26*, 323-332.

[251] B. Weyerich, J. Brunner-Popela, O. Glatter, Small-angle scattering of interacting particles. II. Generalized indirect Fourier transformation under consideration of the effective structure factor for polydisperse systems, *J. Appl. Cryst.* 1999, 32, 197-209.

[252] J. Brunner-Popela, R. Mittelbach, R. Strey, K.-V. Schubert, E. W. Kaler, O. Glatter, Small-angle scattering of interacting particles. III. D2O-C12E5 mixtures and microemulsions with n-octane, *J. Chem. Phys.* **1999**, *110*, 10623-10632.

[253] L. Arleth, J. S. Pederson, Droplet polydispersity and shape fluctuations in AOT [bis(2-ethylhexyl)sulfosuccinate sodium salt] microemulsions studied by contrast variation small-angle neutron scattering, *Phys. Rev. E* **2001**, *63*, 061406-1-061406-18.

[254] T. N. de Castro Dantas, A. C. da Silva, A. A. D. Neto, New microemulsion systems using diesel and vegetable oils, *Fuel* **2001**, *80*, 75-81

[255] J. Silva, M. E. D. Zaniquelli, W. Loh, Light-scattering investigation on microemulsion formation in misture of diesel oil (or hydrocarbons) + ethanol + additives, *Energy Fuels*, **2007**, *21*, 222-226.

[256] E. A. Hazbun, S. G. Schon, R. A. Grey, Microemulsion fuel system, United States Patent 4744796, 1988.

[257] F. Ma, M. A. Hanna, Biodiesel production: a review, *Bioresource Technology*, **1999**, *70*, 1-15.

[258] C. A. W. Allen, K. C. Watts, R. G. Ackman, M. J. Pegg, Predicting the viscosity of biodiesel fuels from their fatty acid ester composition, *Fuel*, **1999**, *78*, 1319-1326.

[259] G. Knothe, Dependence of biodiesel fuel properties on the structure of fatty acid alkyl esters, *Fuel Process Technol.* **2005**, *86*, 1059-1070.

[260] J. Hu, Z. Du, Z. Tang, E. Min, Study on the solvent power of a new green solvent: biodiesel, *Ind. Eng. Chem. Res.* **2004**, *43*, 7928-7931.

[261] S. Wellert, M. Karg, H. Imhof, A. Steppin, H.-J. Altmann, M. Dolle, A. Richardt, B. Tiersch, J. Koetz, A. Lapp, T. Hellweg, Structure of biodiesel based bicontinuous microemulsions for environmentally compatible decontamination: a small angle neutron scattering and freeze fracture electron microscopi study, *J. Colloid Interface Sci.* **2008**, *325*, 250-258.

[262] O. Zech, M. Kellermeier, S. Thomaier, E. Maurer, R. Klein, C. Schreiner, W. Kunz, Alkali oligoether carboxylates - a new class of ionic liquids, *Chem. Eur. J.* **2009**, *15*, 1341-1345.

[263] A. Stoppa, O. Zech, W. Kunz, R. Buchner, The conductivity of imidazolium-based ionic liquids from (-35 to 195)°C. B. Variation of the cation's *N*-alkyl chain. *J. Chem. Eng. Data* submitted.

[264] T. Nishida, Y. Tashiro, M. Yamamoto, Physical and electrochemical properties of 1-alkyl-3-methylimidazolium tetrafuoroborate for electrolyte, *J. Fluorine Chem.* **2003**, *120*, 135-141.

[265] W. Liu, T. Zhao, Y. Zhang, H. Wang, M. Yu, The physical properties of aqueous solutions of the ionic liquid [BMIM][BF$_4$], *J. Solution Chem.* **2006**, *35*, 1337-1346.

[266] J. Vila, L. M. Varela, O. Cabeza, Cation and anion sizes infuence in the temperature dependence of the electrical conductivity in nine imidazolium based ionic liquids, *Electrochim. Acta* **2007**, *52*, 7413-7417.

[267] W.S. Rees Jr., D.A. Moreno, Preparation of monomeric Ba[O(CH$_2$CH$_2$O)$_n$CH$_3$]$_2$ (n = 2, 3), ambient temperature liquid barium compounds, *J. Chem. Soc., Chem. Commun.* **1991**, 1759-1760.

[268] C. J. Pedersen, H. K. Frensdorff, Macrocyclic polyethers and their complexes, *Angew. Chem. Int. Ed.* **1972**, *11*, 16-25.

[269] A. M. Bahl, S. Krishnaswamy, N. G. Massand, D. J. Burkey, T. P. Hanusa, Heavy alkaline-earth polyether carboxylates. The crystal structure of {Ca[OOC(CH$_2$)O(CH$_2$)$_2$]$_2$ O(H$_2$O)$_2$}$_2$, *Inorg. Chem.*, **1997**, *36*, 5413-5415.

[270] A.W. Apblett, J. C. Long, E. H. Walker, Metal organic precursors for yttria, *Phosphorus, Sulfur, and Silicon* **1994**, *93-94*, 481-482.

[271] E. Justus, K. Rischka, J. F. Wishart, K. Werner, D. Gabel, Trialkylammoniododecaborates: anions for ionic liquids with potassium, lithium and protons as cations, *Chem. Eur. J.* **2008**, *14*, 1918- 1923.

[272] J. Fraga-Dubreuil, M.-H. Famelart, J. P. Bazureau, Ecofriendly Fast synthesis of hydrophilic poly(ethyleneglycol)-ionic liquid matrices for liquid-phase organic synthesis, *Org. Process. Res. Develop.* **2002**, *6*, 374-378.

[273] L. Branco, J. N. Rosa, J. J. M. Ramos, C. A. M. Alfonso, Preparation and characterization of new room temperature ionic liquids, *Chem. Eur. J.* **2002**, *8*, 3671-3677.

[274] J. Pernak, K. Sobaszkiewicz, J. Foksowicz-Flaczyk, Ionic liquids with symmetrical dialkoxymethyl-substituted imidazolium cations, *Chem. Eur. J.* **2004**, *10*, 3479-3485.

[275] Z. Fei, W. H. Ang, D. Zhao, R. Scopelliti, E. E. Zvereva, S. A. Katsyuba, P. J. Dyson, Revisiting ether-derivatized imidazolium-based ionic liquids, *J. Phys. Chem. B*, **2007**, *111*, 10095-10108.

[276] K. Izutsu, M. Ito, E. Sarai, Silver-silver cryptate(2,2) ion electrode as a reference electrode in nonaqueous solvents, *Anal. Sci.*, **1985**, *1*, 341-344.

[277] S. Zhang, N. Sun, X. He, X. Lu, X. Zhang, Physical properties of ionic liquids: database and evaluation, *J. Phys. Chem. Ref. Data* **2006**, *35*, 1475-1517.

[278] T. Mosmann, Rapid colorimetric assay for cellular growth and survival: application to proliferation and cytotoxicity assays, *J. Immunol. Meth.* **1983**, *65*, 55-63.

[279] E. M. Eyring, S. Petrucci, M. Xu, L. J. Rodriguez, D.C. Cobranchi, M. Masiker, P. Firman, Lithium ion complexation kinetics by cyclic and acyclic polyethers, *Pure Appl. Chem.* **1990**, *62*, 2237-2241.

[280] J. Grobelny, M. Sokól, Z.J. Jedlinski, Complexation of potassium cations by tetraglyme and 18-crown-6 as evidenced by ^{39}K NMR spectroscopy, *Mag. Res. Chem.* **1991**, *29*, 679-680.

[281] K. Hayamizu, E. Akiba, T. Toshinori, Y. Aihara, ^{1}H, ^{7}Li, and ^{19}F nuclear magnetic resonance and ionic conductivity studies for liquid electrolytes composed of glymes and polyethylene glycol dimethyl ethers of $CH_3O(CH_2CH_2O)_nCH_3$ (n = 3-50) doped with $LiN(SO_2CF_3)_2$, *J. Chem. Phys.* **2002**, *117*, 5929-5939.

[282] P. Johansson, J. Tegenfeldt, J. Lindgren, Modelling amorphous lithium salt-PEO polymer electrolytes: ab initio calculations of lithium ion-tetra-, penta- and hexaglyme complexes, *Polymer* **1999**, *40*, 4399-4406.

Die VDM Verlagsservicegesellschaft sucht für wissenschaftliche Verlage abgeschlossene und herausragende

Dissertationen, Habilitationen, Diplomarbeiten, Master Theses, Magisterarbeiten usw.

für die kostenlose Publikation als Fachbuch.

Sie verfügen über eine Arbeit, die hohen inhaltlichen und formalen Ansprüchen genügt, und haben Interesse an einer honorarvergüteten Publikation?

Dann senden Sie bitte erste Informationen über sich und Ihre Arbeit per Email an *info@vdm-vsg.de*.

Sie erhalten kurzfristig unser Feedback!

VDM Verlagsservicegesellschaft mbH
Dudweiler Landstr. 99   Telefon +49 681 3720 174
D - 66123 Saarbrücken   Fax   +49 681 3720 1749
www.vdm-vsg.de

Die VDM Verlagsservicegesellschaft mbH vertritt

MIX
Papier aus verantwortungsvollen Quellen
Paper from responsible sources
FSC® C105338

Printed by Books on Demand GmbH, Norderstedt / Germany